DESIGN AND CONSTRUCTION OF LABORATORY GAS PIPELINES

A Practical Reference for Engineers and Professionals

DESIGN AND CONSTRUCTION OF LABORATORY GAS PIPELINES

A Practical Reference for Engineers and Professionals

James Moody

AAP | APPLE ACADEMIC PRESS

Apple Academic Press Inc.	Apple Academic Press Inc.
3333 Mistwell Crescent	1265 Goldenrod Circle NE
Oakville, ON L6L 0A2	Palm Bay, Florida 32905
Canada USA	USA

© 2019 by Apple Academic Press, Inc.

First issued in paperback 2021

Exclusive worldwide distribution by CRC Press, a member of Taylor & Francis Group

No claim to original U.S. Government works

ISBN 13: 978-1-77463-414-1 (pbk)
ISBN 13: 978-1-77188-723-6 (hbk)

Library and Archives Canada Cataloguing in Publication

Title: Design and construction of laboratory gas pipelines : a practical reference for engineers and professionals / James Moody.

Names: Moody, James, 1950- author.

Description: Includes bibliographical references and index.

Identifiers: Canadiana (print) 20190063041 | Canadiana (ebook) 20190063114 | ISBN 9781771887236 (hardcover) | ISBN 9780429469732 (PDF)

Subjects: Laboratories—Equipment and supplies. | LCSH: Gas pipelines—Equipment and supplies. | LCSH: Gas pipelines—Design and construction. | LCSH: Gases.

Classification: LCC Q183.A1 .M66 2019 | DDC 665.7/44—dc23

CIP data on file with US Library of Congress

Apple Academic Press also publishes its books in a variety of electronic formats. Some content that appears in print may not be available in electronic format. For information about Apple Academic Press products, visit our website at **www.appleacademicpress.com** and the CRC Press website at **www.crcpress.com**

ABOUT THE AUTHOR

James Moody

James (Jim) Moody has more than 40 years experience working continuously in the field of piped gases, particularly in the healthcare industry and for industrial, scientific, and university laboratories.

James's experience includes working as a Project Manager and Design Engineer on multimillion-dollar projects and as the Managing Director of one of the largest medical and laboratory gas installation companies in Australia. James is currently working as a consulting engineer specializing in medical and laboratory gas systems.

Over the years, James has been involved in the design and development of products and equipment associated with the controls and alarms for gas reticulation systems, winning the Australian Design Award for his tourniquet control panel used in operating theatres. Another design breakthrough has been in plume (smoke) evacuation systems for the removal of the plume generated by medical devices such as YAG Lasers and devices for processes in electrocautery and diathermy.

James has been a member of Standards Australia (HE–017 Medical Gas Systems) and a member of Australia's delegation member of the International Organization for Standardization (ISO), since the 1980s. The HE–017 Medical Gas Systems Committee manages a wide variety of standards publications, including AS 2896–2011 relating to medical gas pipeline systems. In 2015, James was elected as the chair of the committee. As a member of the ISO committee, for six years James was involved in the writing of the *ISO 16571 Systems for Evacuation of Plume* standard for the medical devices. James has also written several papers on piped medical and laboratory gases.

CONTENTS

ABBREVIATIONS

A/V	audible/visual
AAS	atomic absorption spectrophotometer
ANSI	American National Standard Institute
Ar	argon
AS	Australian Standard
ASTM	American Society of Testing Materials
BESA	British Engineering Standards Association (BSI)
BMS	Building Management System
BS	British Standard
BSI	British Standards Institution
C_2H_2	acetylene
CAN/CSA	Canadian Standards Association
CE	Confirmité Européen/European Conformity (CE Marking)
CENELEC	European Committee for Electrotechnical Standardization/ Comité Européen de Normalisation Electrotechnique/ Europäisches Komitee für Elektrotechnische Normung
CGA	Compressed Gas Association
CH4	methane
CO	carbon monoxide
CO2	carbon dioxide
DEM	door entrance module
DIN	German Standard (Deutsches Institut für Normung)
EN	European Standard
EPR	electron paramagnetic resonance
FBA	flashback arrestor
FFL	finished floor level
FID	flame ionization detectors
FPD	flame photometric detectors
GC	gas chromatograph
GC-MS	gas chromatograph mass spectrometer
GFAAS	graphite furnace atomic absorption spectrometry
H_2	hydrogen
He	helium

HLI	high-level interface
HP	high purity
hPa	hectopascals
HPLGC	high-pressure liquid-gas chromatograph
HTM	hospital technical memorandum
ICP	inductive coupled plasma
ID	inside diameter
IEC	International Electrotechnical Commission
ISA	International Society of Automation
ISO	International Standards Organization
IVF	in vitro fertilization
kPa	kilopascals
kPaA	kilopascals absolute
LC	liquid chromatography
LCGC	liquid coupled gas chromatograph
LCMS	liquid chromatography-mass spectrometer
LEL	lower explosive limit
LN2	liquid nitrogen
LPG	liquefied petroleum gas
MBE	molecular beam epitaxy
MRI	magnetic resonance imaging
MS	mass spectrometer
MSDS	material safety data sheet
N2	nitrogen
N2O	nitrous oxide
NASA	National Aeronautics and Space Administration
NFPA	National Fire Protection Authority
NMR	nuclear magnetic resonance spectroscopy
NPT	normal temperature and pressure
NRV	non-return valves
NZS	New Zealand Standards
O2	oxygen
OD	outside diameter
OHS	occupational health and safety
P10	gas mixture, 10% methane in argon
PCI	pulverized coal injection
PFD	probability of failure on demand
PFH	probability of failure per hour

ppb	parts per billion
ppm	parts per million
PSA	pressure swing adsorption
psi	pounds per square inch
PVDF	polyvinylidene fluoride
QuEChERS	quick easy cheap effective rugged safe
SCC	structural corrosion cracking
SEMI	Semiconductor Equipment and Materials International
SIF	safety integrated function
SIL	safety integrity level
SIS	safety instrumented system
STEL	short-term exposure level
TWA	time-weighted average
UEL	upper explosive limit
UHP	ultra-high purity
UL	Underwriters Laboratories
UPS	uninterruptible power supply
VIE	vacuum insulated evaporator
VIV	vacuum insulated vessel
VOC	volatile organic compound
VSD	variable speed drive
XRF	x-ray fluorescence

FOREWORD

A book of this type is a long overdue and finally fills a void that has been present in this industry for many decades. It deals with the design and safe installation of different gas systems and is written in plain English suitable for both the novice or expert. Importantly it is written in texbook style where each chapter is complete by itself in dealing with each specific gas and thereby not necessitating that each reader read the entire document.

In my personal experience, if a book like this had existed three decades ago, my own work would have been much improved. I was setting up a Gene Targeting and Stem Cell Laboratory at the John Curtin School of Medical Research at the Australian National University. This involves the genetic manipulation of mouse embryonic stem cells and their microinjection into mouse pre-implantation embryos, all done under CO_2 buffered tissue culture. Our initial results were very poor even though we were using state-of-the-art equipment and reagents. By chance, Jim Moody visited my laboratory and immediately pointed out a major flaw in our laboratory design: we had used copper to pipe the CO_2 gas. Jim directed me to an article in *Nature* by Brinster in 1972 that clearly shows the toxic effect of copper ions on pre-implantation embryos. We immediately re-piped our CO_2 in stainless steel, and overnight our success dramatically improved to become one of the most successful gene targeting laboratories both in Australia and internationally.

It is this type of personal knowledge that this book describes that has been gathered over a lifetime career. I cannot recommend this book more highly for those of us that always want to get things perfectly correct.

—Professor Klaus Matthaei BSc Hons (UNSW), PhD (ANU)
Head, (now retired)
Gene Targeting and Stem Cell Laboratory,
The John Curtin School of Medical Research,
The Australian National University,
Canberra, Australian Capital Territory 0200, Australia

ACKNOWLEDGMENTS

ALVI Technologies Pty Ltd
Unit 2/79 Station Road,
Seven Hills NSW 2147,
Australia,
Tel: +61-2-9838-7220,
www.alvi.com.au; www.gasalarm.com.au

ASTM International
#100, Barr Harbor Drive,
PO Box C700, West Conshohocken,
PA 19428-2959, USA,
Tel: +1-610-832-9634,
http://www.astm.org

Atlas Copco Australia Pty Ltd
Atlas Copco Compressors (Australia),
#3, Bessemer Street,
Blacktown, NSW 2148,
Australia,
Tel: +61-2-96219795,
http://www.atlascopco.com.au

Busch Australia Pty Ltd
#30, Lakeside Drive,
Broadmeadows VIC 3047,
Australia,
Tel: +61-1800-639-087,
https://www.buschvacuum.com/au/en

Cryoquip Australia
#14, Zenith Road,
Dandenong, VIC 3175,
Australia,
Tel: +61-(0)-402-302-251; +61-(0)-3-9791-7888,
http://www.cryoquip.com.au/store/index.asp

Düperthal Sicherheitstechnik
Sitz: Karlstein,
Registergericht: Amtsgericht Lemgo HRA 3229,
Pers. haft. Ges.: DÜPERTHAL Beteiligungs GmbH,
Sitz: Oerlinghausen,
Registergericht: Amtsgericht Lemgo HRB 4193,
Frankenstrasse 3,
63791 Karlstein,
Germany,
http://www.dueperthal.com

Gascon Systems Pty Ltd
24, Ford Crescent, Thornbury VIC 3071,
Australia;
PO Box 284, Balwyn North VIC 3104,
Australia,
Tel: +61-3-9499-4100,
http://www.gascon.com.au

MSR Group
Würdinger Str. 27A,
94060, Pocking,
Germany,
Tel: +49-8531-90040,
http://www.msr-electronic.de/en

SAI Global
Level 37, 680 George Street,
Sydney NSW 2000,
Australia,
Tel: +61-2-8206-6742,
http://www.saiglobal.com

Veeco Corporate Headquarters
MBE Business Unit,
#1, Terminal Drive,
Plainview, NY 11803,
USA,
Tel: +1-(516)-677-0200,
http://www.veeco.com/mbe

INTRODUCTION

The specific information regarding the design and installation of laboratory gas pipelines that is essential for the preparation of a competent pipeline design is found in a variety of specialist reference texts, manufacturer's literature or in a variety of university publications that under normal circumstances are widely spread and not easily accessible.

Hopefully, this book will provide sufficient references and the basic information necessary to instigate the search for the information required to allow design engineers, laboratory managers, and installation technicians to source the information they require for their own information or to provide their clients with a laboratory gas system suitable for the application.

The current use of specifications predominantly taken from medical gas standards for this type of work is not always suitable, these standards are for use with medical grade gases that have a purity level of 99.5% which is well below the levels required in laboratories that start at 99.9% for general industrial use through to 99.9995% (Ultra High Purity (UHP)) and higher. The medical gas standards are also unsuitable for use with the oxidizing, flammable, and, in some instances toxic gases that are regularly encountered in laboratories.

As gas purity increases, so the methodology used to design a piping system must vary to meet those parameters. A design using a medical standard should be used as a guide that follows the laboratory gas need only be 99.5% pure. This is effectively an industrial purity grade gas which is well below the generally accepted minimum for laboratory use.

The book is arranged into seven chapters:

Chapter 1 – *Laboratory Gases, Types, and Equipment Encountered* – provides information on types of laboratories that use laboratory gases and the equipment used for the various piped gases.

Chapter 2 – *Laboratory Gas Supply: Plant and Equipment* – examines the equipment used to supply gases for laboratory use and also the designs for the plant used for in-house air and vacuum systems.

Chapter 3 – *Laboratory Gas Pipeline Construction* – provides details on the various methods of construction and the materials used to ensure the purity of the gases remain as supplied from the manufacturers.

Chapter 4 – *Laboratory Gas Pipeline Design* – incorporates the design methodology used to meet the various requirements of the laboratory and the information required to ensure the correct engineering is provided.

Chapter 5 – Gas Data: Inert Gases – is a gas database that provides information on the purity levels of the inert gases and the data on the equipment used for pipelines and compatibility issues for the gases used in laboratories.

Chapter 6 – *Gas Data: Flammable and Toxic Gases* – is a database that provides information on the purity levels of the flammable and toxic gases and the data on the equipment used for pipelines and compatibility issues for the gases used in laboratories.

Chapter 7 – *Sample Specification* – is an example only of a simple laboratory gas specification that provides guidelines on the information necessary to provide a set of design documents. This is not for use as there is considerably more information that must be included to meet the laboratory-specific requirements.

There is one point that cannot be stressed sufficiently; while the information provided here will assist the design engineer to draft a viable working system design and specification it cannot, however, be used as the complete resource for a laboratory gas pipeline system. The engineer will require additional gas specific information from the proposed laboratory to confirm pressures, flows, and purities for each terminal connection. In addition, the equipment manufacturers must be contacted to discuss the analytical equipment that is to be connected to the gas pipelines, without which the design cannot meet the diverse requirements for any scientific laboratory.

CHAPTER 1

LABORATORY GASES, TYPES, AND EQUIPMENT ENCOUNTERED

1.1 INTRODUCTION

The laboratory equipment that uses the gases piped through the laboratories is specific in its demands for UHP gas supplies to enable the sampling procedures to operate correctly and on a repetitive basis. Any impurity in the gas stream can provide incorrect or false readings and thereby render any testing procedure unusable.

The equipment that uses these gases is highly accurate, and the results that are being provided can be the result of years of research or maybe in a laboratory that is testing clinically important samples that may provide lifesaving outcomes.

This equipment is usually installed long after the pipelines have been installed, tested, and handed over to the laboratory; and the laboratory gas system design engineer may have little knowledge of the final use of the pipelines that they've designed.

1.2 LABORATORY GASES

The laboratory gases that are referenced in this book are limited to those most commonly encountered in the pipeline installation industry. There are many gas mixtures and rare gases that are likely to be encountered especially in some very specialized research laboratories, e.g., krypton and xenon, which would require special consideration and extensive research about the properties of these gases prior to designing a pipeline and we do not propose to provide information on these here.

However, we have included the information on the following gases:

- Acetylene;
- Instrument air;
- Argon;
- Argon (liquid);
- Carbon dioxide;
- Carbon dioxide (high-pressure liquid);
- Carbon monoxide;
- Helium;
- Helium (liquid);
- Hydrogen;
- Methane;
- Methane in Argon (P10);
- Nitrogen (gaseous to 1,000 kPa);
- Nitrogen (liquid);
- Nitrous oxide;
- Oxygen (gaseous);
- Oxygen (liquid); and
- Vacuum.

Note: Even though vacuum is not actually a gas, it has been included here since it is regularly included as part of most laboratory gas pipeline installations. Vacuum pipelines are used in a variety of application that requires a range of pressures and flows, each of which requires different plant and equipment; especially, where the levels of vacuum required to evaporate some VOC's is necessary.

We have included gas specific information in Chapters 5 and 6 with some limited information on the materials of construction for each gas type. Detailed information on the materials and engineering design used can also be found in Chapters 3 and 4. And, an example of a design specification is shown in Chapter 7.

1.3   LABORATORY TYPES

There are a wide variety of laboratories that use piped laboratory gases. These range from simple school laboratories using liquefied petroleum gas (LPG) for Bunsen burners to sophisticated teaching and research laboratories in universities and public and privately funded research and scientific facilities.

Some regularly encountered laboratory types that have piped gases include:

- Pathology Laboratories;
- Hospital-Based Clinical Research Laboratories;
- University Research and Teaching Facilities;
- Scientific Testing Laboratories; and
- IVF Laboratories.

Each laboratory type has different gas requirements to suit the equipment they propose to use. In many instances, these laboratories will use a mixture of similar equipment; whereas, some specialized facilities may include only one or two facilities-focused items.

In each location, it is necessary to ensure that the design engineer is fully conversant with the requirements of the laboratory and the uses of the specific gases and the format in which each gas will need to be supplied.

1.4 LABORATORY EQUIPMENT

The common types of equipment that are usually encountered in most laboratories include:

1. *Atomic Absorption Spectrophotometer (AAS) (Figure 1.1)*

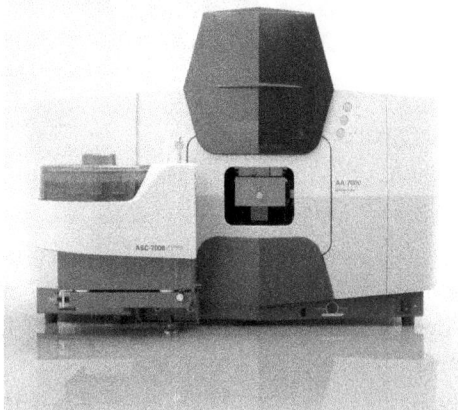

FIGURE 1.1 Shimadzu model ASC-7000 Atomic Absorption Spectrophotometer.
(**Source:** Shimadzu Australasia. With permission).

The gases required here include:

- Acetylene;
- Nitrous oxide (optional depending on the laboratory); and
- Instrument air (optional depending on the laboratory).

2. Excimer Lasers

The information and pictures in this section are provided with the kind permission of Dr. Ralph Delmdahl, Coherent Laser Systems GmbH & Co. KG (Figure 1.2).

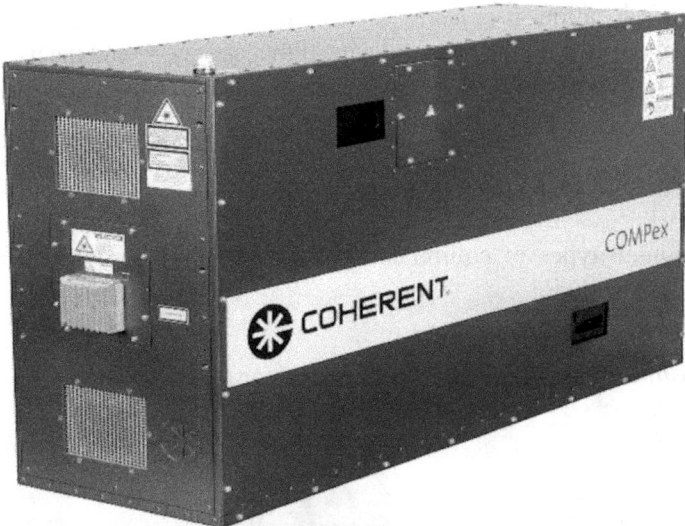

FIGURE 1.2   Coherent COMPex excimer laser.

(**Source:** Dr. Ralph Delmdahl, Coherent Laser Systems GmbH & Co. KG. With permission.)

An excimer laser can be described as an ultraviolet laser and is used in the manufacture of microelectronic chips and semiconductor-based integrated circuits, in deep-ultraviolet photolithography and Lasik eye surgery (Figures 1.3 and 1.4).

The gases required may vary depending on the site-specific requirements. They may also be supplied as a premixed gas specifically for the laboratory. For example,

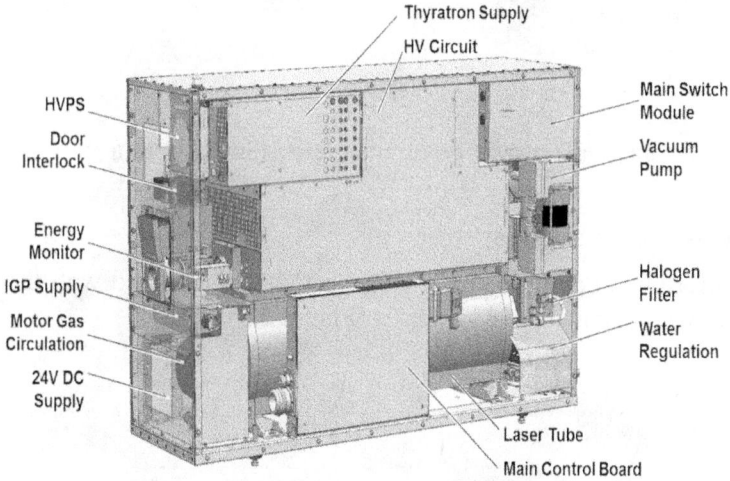

FIGURE 1.3 Cutaway view of the coherent COMPex excimer laser.

(**Source:** Dr. Ralph Delmdahl, Coherent Laser Systems GmbH & Co. KG. With permission.)

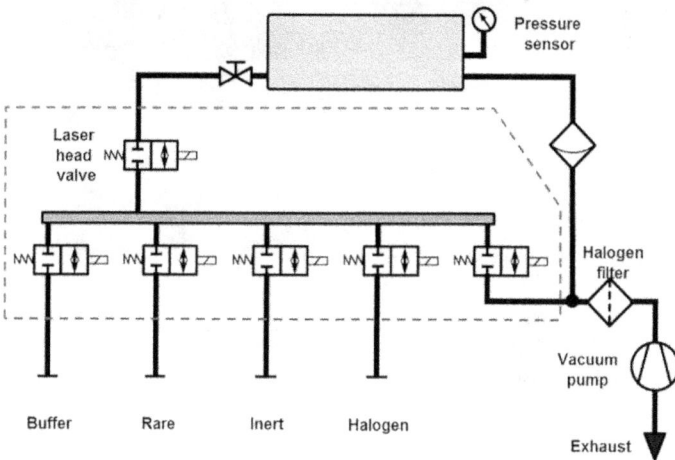

FIGURE 1.4 Gases schematic for the coherent COMPex excimer laser.

(**Source:** Dr. Ralph Delmdahl, Coherent Laser Systems GmbH & Co. KG. With permission.)

- Fluorine;
- Argon;
- Krypton;

- Instrument air;
- Xenon; and
- Helium.

Note: The above gases may be supplied individually or as a premixed gas.

3. Gas Chromatographs

The information and pictures in this section are provided with the kind permission of Shimadzu Australasia (Figure 1.5).

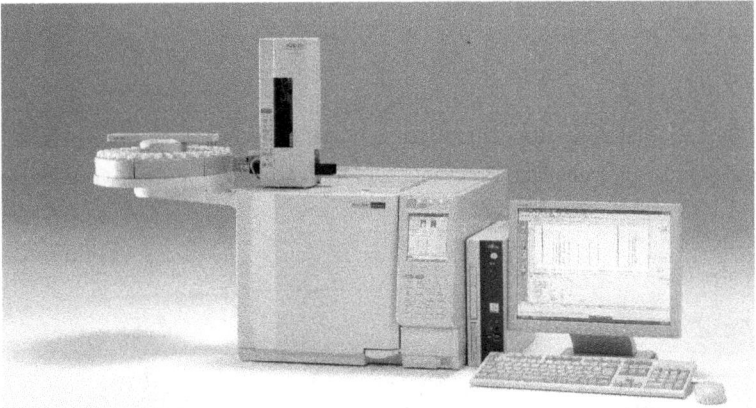

FIGURE 1.5   Shimadzu gas chromatograph.
(**Source:** Shimadzu Australasia. With permission.)

A gas chromatograph (GC) is an analytical instrument used to analyze vaporized compounds without destroying the sample. The GC uses two phases, a mobile and a stationary phase. The mobile phase is the carrier gas, usually an inert gas such as helium or nitrogen, and the stationary phase is a microscopic layer of liquid on an inert solid support. The stationary phase lines the inside surface of a glass or metal tube known as a column.

There are two types of columns are used in GC. They are:

a.   Capillary columns have a very small internal diameter of a few tenths of a millimeter. The column walls are coated with the stationary phase. Most capillary columns are made of fused silica with a polyimide outer coating. These columns are flexible, so a very long column can be wound into a small coil.

b. Packed columns contain a finely divided, inert, solid support material (e.g., diatomaceous earth) coated with a liquid or solid stationary phase. The nature of the coating material determines what type of materials will be most strongly adsorbed. There are numerous columns available that are designed to separate specific types of compounds. Most packed columns are 1.5–10 m in length and have an internal diameter of 2–4 mm. The outer tubing is usually made of stainless steel or glass.

GC has measurement accuracy levels of parts per billion concentration. The gases required include:

- Nitrogen;
- Hydrogen;
- Argon;
- Helium; and
- Carbon dioxide (high pressure).

4. Inductively Coupled Plasma (ICP) (Figure 1.6)

FIGURE 1.6 Shimadzu inductively coupled plasma.
(**Source:** Shimadzu Australasia. With permission.)

The gases required include:

• Argon; and
• Instrument air (optional depending on the laboratory).

5. Molecular Beam Epitaxy

Molecular-beam epitaxy (MBE) is used to deposit a thin film of single crystals and is used in the manufacturing process of semiconductor materials and in nanotechnology.

The following information is provided with the permission of Veeco (Figure 1.7):

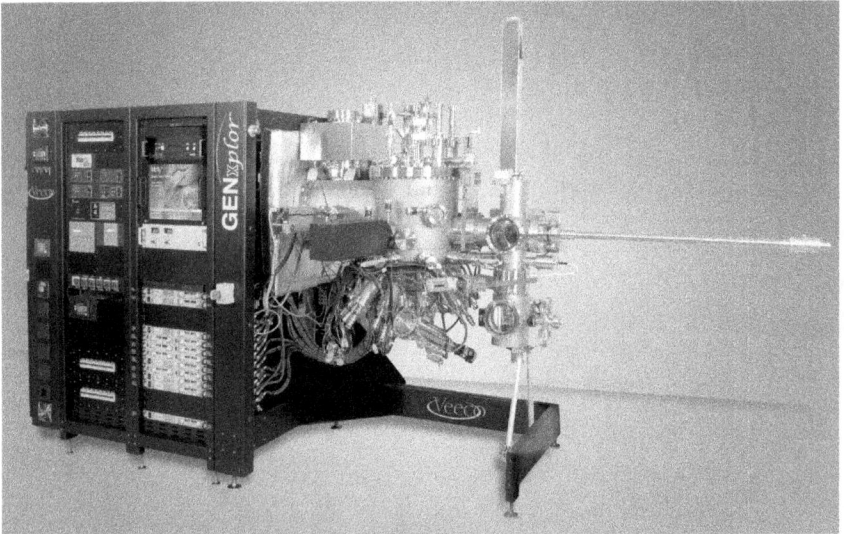

FIGURE 1.7   Veeco molecular beam epitaxy.
(**Source:** Information courtesy of Veeco. With permission.)

Veeco's award-winning GENxplor™ R&D MBE System uses Veeco's proven GEN10™ MBE system growth chamber design and features unmatched process flexibility, perfect for materials research on emerging technologies such as UV LEDs, high-efficiency solar cells and high-temperature superconductors. Its efficient single frame design combines all vacuum hardware with onboard electronics to make it up to 40% smaller than other MBE systems, saving valuable lab space.

Because the manual system is integrated on a single frame, installation time is reduced. The open architecture design of the GENxplor MBE system also improves ease-of-use, provides convenient access to effusion cells and easier serviceability when compared to other MBE systems. When coupled with Veeco's recently-introduced retractable sources, the GENxplor MBE system represents the state-of-the-art in oxide materials research.

- High quality epitaxial layers on substrates up to 3″ in diameter.
- Unique, single frame architecture improves ease-of-use, provides convenient source access and enhanced serviceability.
- Efficient, all-in-one design combines the manual system with onboard electronics for 40% of lab space savings compared to other MBE systems.
- Ideal for cutting-edge research on a wide variety of materials including GaAs, nitrides, and oxides.
- Molly® software integrates easy recipe writing, automated growth control, and always-on data recording.
- Optional Nova™ ultra-high temperature substrate heater for proven performance at 1850°.
- Direct scalability to GEN20™, GEN200®, and GEN2000® MBE systems".

The gases required are:

- Liquid nitrogen;
- Gaseous nitrogen; and
- High vacuum (10-8-10-12 Torr) insulated pipelines using a combination of static and dynamic vacuum insulated pipework.

1.4.1   REACTORS AND CALORIMETERS

The information and pictures in this section are provided with the kind permission of Parr Instrument Company, Moline Illinois, USA.

Parr Instrument Company manufactures pressure vessels and stirred reactors for use in the laboratory study of high-pressure chemistry around the world. Typical research includes investigation of novel uses of existing compounds as well as novel methods by which existing compounds can be manufactured. These research efforts may take place in academic, governmental, or industrial laboratories in businesses such as petroleum, petrochemical, chemical, polymer, pharmaceutical, biofuels, plant extractions, and hydrometallurgy (Figure 1.8).

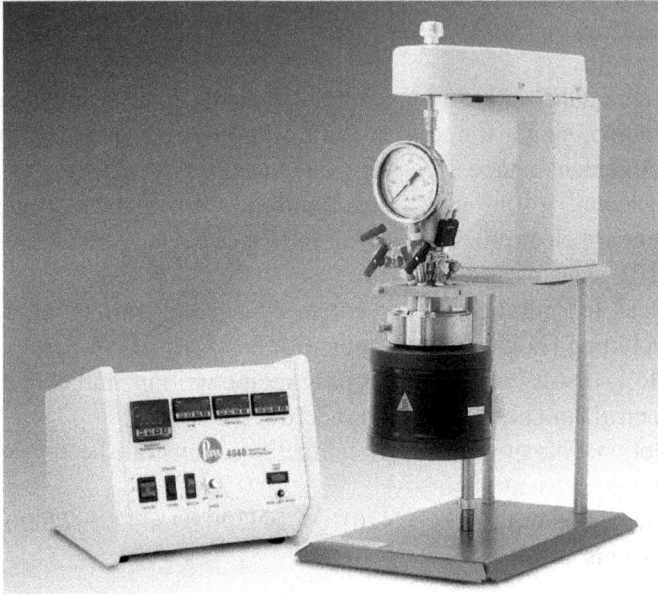

FIGURE 1.8   Parr model 4848 calorimeter.

- The most common model of the stirred reactor is the 300 mL Parr Mini-Reactor. This is the workhorse of many laboratories. When made of stainless steel, it is designed for use to 350°C at pressures as high as 20,000 kPa (3,000 psi).
- Another popular series is the 1-L and 2-L reactors that are rated for use to 350°C at pressures as high as 13,000 kPa (1,900 psi). These reactors are available in Bench-top or Floor-stand mounting (Figures 1.9a and 1.9b).
- The largest standard Parr reactor has a capacity of 19 liters (5-US gallons) and is designed and rated for use to 13,000 kPa (1,900 psi) at 350°C (Figure 1.10).

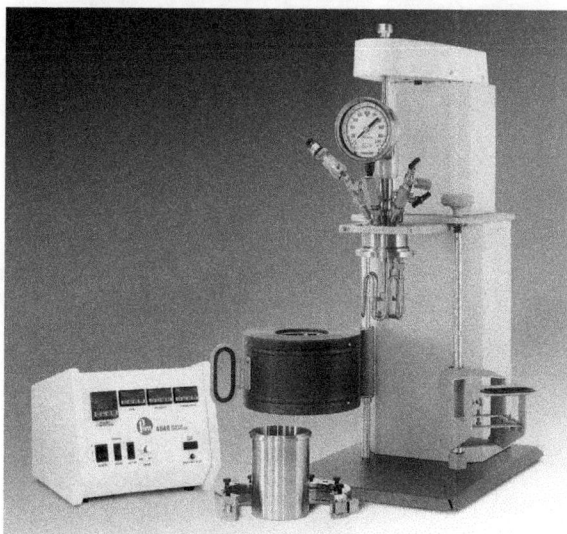

FIGURE 1.9a Parr bench mounted models 1–L and 2-L calorimeters.

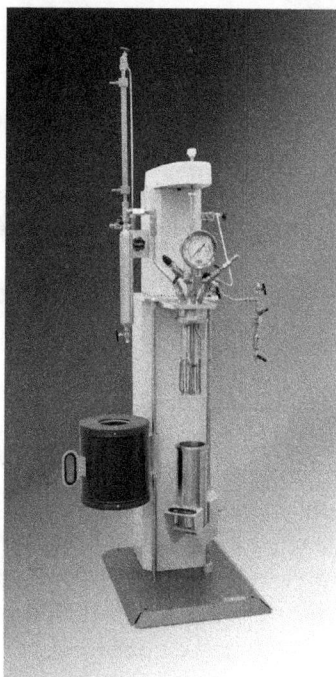

FIGURE 1.9b Parr floor mounted models 1–L and 2-L calorimeters.

FIGURE 1.10   Parr floor mounted model calorimeter.

- The heads of most reactors are equipped with a gas release valve to reduce internal pressure, a liquid sampling valve and a gas inlet valve. Fittings on the head also include a pressure gage, an internal cooling coil, and a safety rupture disc. Figure 1.11 shows a typical arrangement of it.
- It is common that the gas release valve and the safety rupture disc are piped to a safe outlet for collection or proper disposal. Although heating of the vessel contents will increase the internal pressure, the initial pressure charge most often comes from a cylinder of compressed gas. These gas cylinders are typically supplied with a pressure of approximately 15,000 kPa (2,200 psi). A pressure reducing regulator and tubing for gas transport are required for connection to the reactor's gas inlet valve.
- Another series of vessels available from Parr Instrument Company is designed for use to 34,500 kPa (5,000 psi) at temperatures to 500°C. These vessels often require the use of a gas compressor to increase typical tank pressures to achieve the full operating capabilities of these reactors (Figure 1.12).

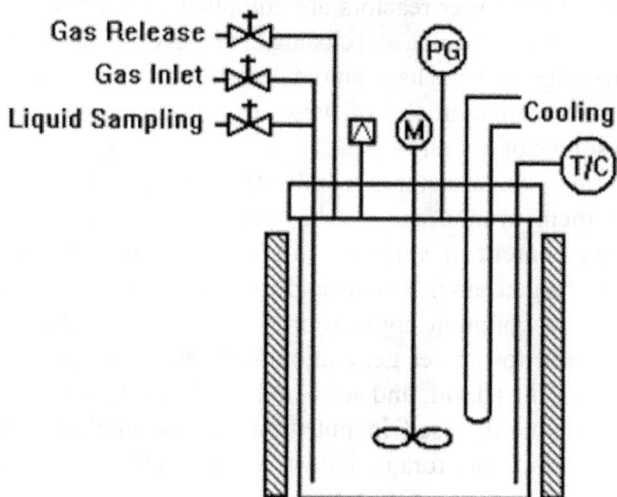

FIGURE 1.11 Typical gas connections schematic.

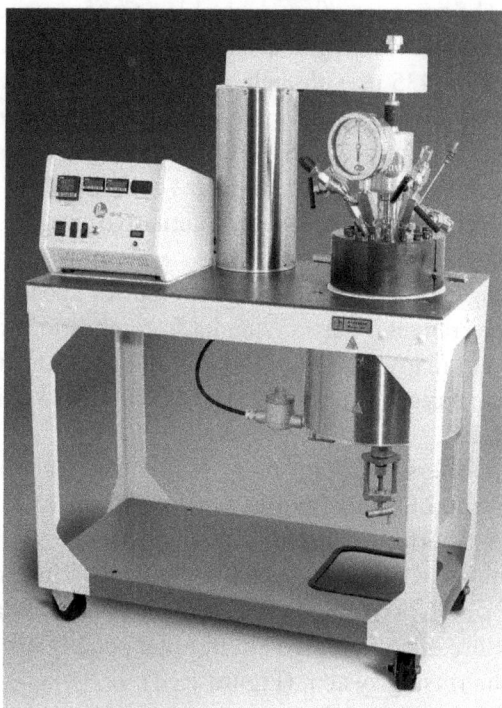

FIGURE 1.12 Parr model 4848 calorimeter.

- Some of the larger reactors are equipped with a pneumatic lift to assist with the opening and closing of vessels with heavier cylinders. Some other systems have automated shut-off valves and automated pressure regulation. All of these features require a 700 kPa (100 psi) source of dry air.
- Parr also manufactures combustion calorimeters and distributes them around the world. They allow users to study the energy content of various samples, by combusting the sample in an oxygen environment and measuring the energy liberated. The most common application for such calorimeters is testing fuels used for power generation and heating purposes like coal, biomass, diesel, oil, and solid and liquid waste. Calorimeters are also commonly used in nutrition and metabolism studies and in the food and forage industries. Propellants and explosives testing, as well as safety studies also require specialized combustion calorimeters.
- All combustion calorimeters need a source of high-purity (>99.5%) oxygen. The presence of any burnable components (e.g., hydrogen) is not acceptable. Most global standards require an oxygen pressure of 3,000 kPa (435 psi), though occasionally 4,000 kPa (580 psi) is used. Stable oxygen supply pressure is critical. Parr calorimeters are therefore typically delivered with a gas pressure regulator, which can be attached to standard gas cylinders with approximately 15,000 kPa (2,200 psi) pressure. Alternative oxygen supplies can be used if the above criteria are met.
- Parr offers various calorimeters tailored for specific requirements and price ranges. Parr's standard combustion calorimeter as described in many modern standards and books is the Parr 6200. It requires an oxygen supply of only 3,000 kPa (435 psi) for operation (Figure 1.13).
- Parr also offers the Parr 6400 Automatic Calorimeter, specifically designed for laboratories with high sample throughput. This calorimeter performs formerly manual operations like gas filling, gas venting and vessel rinsing automatically. Thus, in addition to a 3,000 kPa (435 psi) oxygen line, it requires a 550 kPa (80 psi) pipeline pressure of compressed air, nitrogen, or any other non-reactive gas for the rinsing system (Figure 1.14).

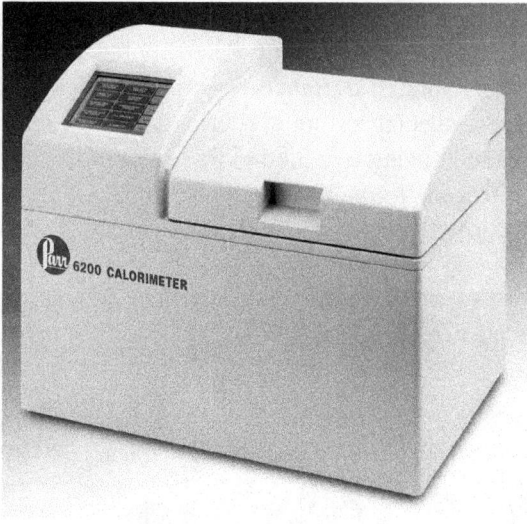

FIGURE 1.13   Parr model 6200 calorimeter.

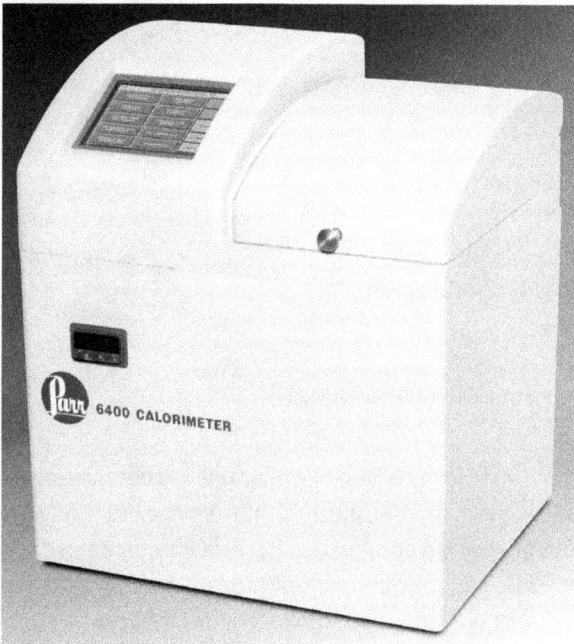

FIGURE 1.14   Parr model 6400 calorimeter.

- For defense applications and other work with samples with extremely high energy content, Parr supplies the 6790 Detonation Calorimeter. Though designed for operation with 3,000 kPa (435 psi) oxygen, special testing scenarios may be required to operate the calorimeter using 0 to 3,000 kPa (435 psi) of other compressed gasses: nitrogen, argon, etc. The gas supply for such special applications is provided by the user directly (Figure 1.15).

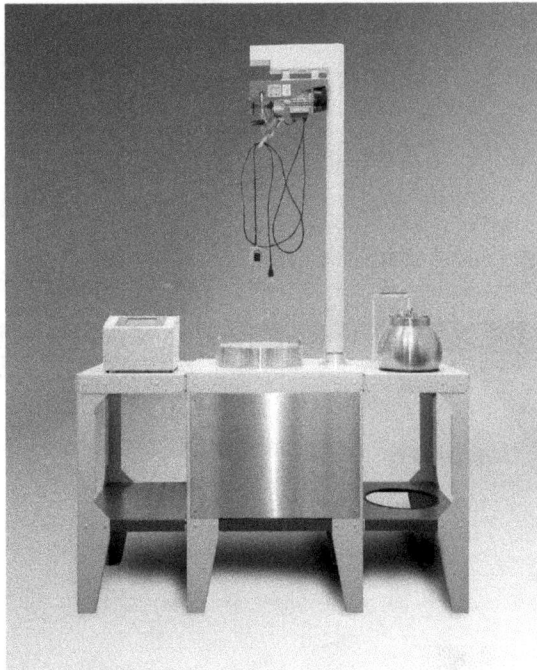

FIGURE 1.15   Parr model 6790 calorimeter.

- The Parr 1901 Oxygen Vessel Apparatus is a complete system designed for the study of combustion within a pressure vessel. The package includes an oxygen combustion vessel (for converting solid and liquid combustible samples into soluble forms for chemical analysis) and a full set of operating accessories. The accessories in 1901 include an oxygen filling connection, ignition unit, ignition cords, vessel head support stand, vessel lifter, combustion capsules, fuse wire, gelatine capsules, spare gaskets, and sealing rings (Figure 1.16).

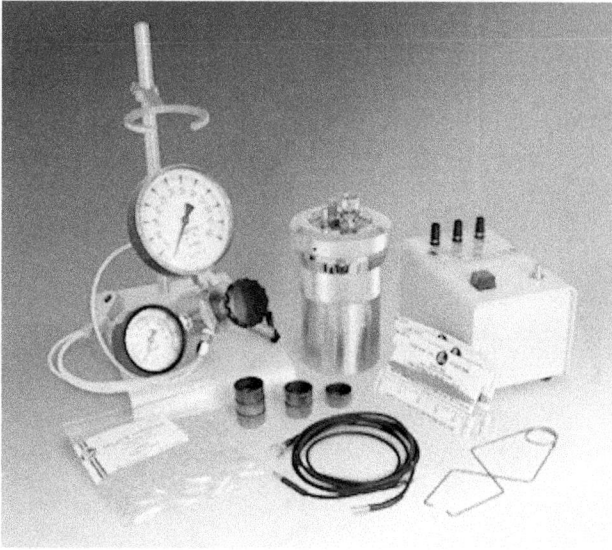

FIGURE 1.16   Parr model 1901 calorimeter.

Note: The above information provides examples of the varying oxygen pressures that are likely to be required for oxygen calorimeters; the range is wide-ranging from 550 kPa (80 psi) to 34,500 kPa (5,000 psi) at temperatures to 500°C. It is absolutely imperative to investigate the equipment requirements prior to providing an oxygen supply. It may also be necessary to provide a 700 kPa (100 psi) source of dry air for controls in some instances.

KEYWORDS

- laboratory equipment
- laboratory gases
- laboratory types
- Parr model calorimeter

CHAPTER 2

LABORATORY GAS SUPPLY: PLANT AND EQUIPMENT

2.1 INTRODUCTION

The selection of the plant that will supply the various gases to the laboratory needs to be targeted at the specific demands of the equipment being used. The variables that must be taken into consideration include the pressure, flow, purity, and the respective items of plant and the equipment that will transport the gases from their sources to the terminal outlets throughout the pipeline system. Any item of plant or pipe in the system could contaminate the gas and render it unacceptable for use. It must be remembered that any impurity in the gas stream will render what may be the results of months of research invalid or incorrect.

There are a variety of ways to store and supply gases that differ from location to location. Each of the sources of supply needs to be selected to take into consideration the location, the distance from the supplier's manufacturing plant, the availability of the gases, the method by which the gases will be delivered to the laboratory from the source, and how the gas supplier will provide the gases.

The sources of the various gases may be from cryogenic vessels, high-pressure cylinders, compressor or vacuum plants, pressure swing adsorption systems or from site manufactured gas equipment. The use of single sources of cryogenic gas supplies is suitable for large high demand facilities. They are also suitable for laboratories that require liquid nitrogen for decanting into portable dewars and may simultaneously provide an alternative gaseous supply for laboratories. The localized plant for instrument air may be an alternative to a cylinder supply and for vacuum; there is no alternative to providing a plant.

Every supply source needs to be individually designed. The pressure requirements of many gases will vary depending on the laboratories requirements and many gases are available in a range of pressures and purities as well as in liquid or gaseous form.

This chapter looks at a number of alternative types of supply. It is the responsibility of the designer to select the most suitable method of providing an appropriate, fit for purpose and cost-effective outcome for the laboratory.

2.2   INSTRUMENT AIR PLANT AND EQUIPMENT

2.2.1   INSTRUMENT AIR COMPRESSORS

There are a variety of compressors in a number of styles and types in use for the production of instrument air for laboratory purposes. The equipment currently available for providing oil-free clean instrument air uses basic principles to compress, cool, dehydrate, and filter to supply a relatively clean and dry instrument air source.

Installations of recent times will consist of a variety of manufacturers as well as different types of compressors that include reciprocating, screw, scroll, and centrifugal compressors. In Australia, the only thing that these units will have in common is that for instrument air use they should be oil-free. At this time there does not appear to be any data that offers definitive information about what types of units to use under what circumstances. With the exception of medical air, the use of instrument air is currently unregulated and designs are left to the system designers.

The following information provides guidelines on the selection of the correct compressor for the particular installation based on the demands of the system from both a cleanliness and capacity perspective. The final selection will be up to the designer and a poor choice at the design stage could render the air supply useless and may even contaminate the pipeline.

To start the selection process, we need to find the basic requirements that the instrument air must meet. The highest levels of purity can be found in high-pressure cylinders provided by the gas companies and are supplied with certification including a complete analysis of the air in the cylinder. The cylinder supply of instrument air is available from purity levels of 99.5% through to 99.99995% or for particularly sensitive equipment cylinders with certification; and a complete analysis of the makeup of the air in the cylinder is necessary. It is not possible for a local compressor plant installed at the laboratory site to provide this quality instrument air.

The selection of the gas purity can only be done by those who will be operating the equipment or from advice obtained from the equipment manufacturer or supplier who will be able to provide information pertaining to the purity levels that their equipment requires to function correctly. The instrument air source can then be selected to suit the system, i.e., compressor plant or cylinder supply or a mixture of both. In this chapter, we are looking at compressor plant supply. So, we will assume that this has been decided upon.

It cannot be stressed sufficiently that the purity of the instrument air will determine the accuracy of the results provided by the laboratory equipment. The instrument air is connected to analytical equipment that may be able to provide information with exacting parameters. However, if the instrument air (or any gas) is not able to meet the purity levels demanded then the results will be worthless.

For each grade of instrument air, the major concerns are the pressure, level of moisture or dew point of the air, soil contamination and any remaining contaminants that may not have been removed during production.

To provide very high purity levels of instrument air, the system requires continuous monitoring and testing to ensure the necessary maximum contamination levels are being maintained.

2.2.2   PURITY LEVELS OF INSTRUMENT AIR

In ISO 8573, parts 1 through 9 [50–58], there are detailed explanations of purity levels and the methods of testing for each of them. ISO 8573 Part 1 [50] categorizes a number of classes each of which has minimum acceptable limits. They are:

- Table 1: Compressed air purity classes for particulates.
- Table 2: Compressed air purity classes for humidity and liquid water.
- Table 3: Compressed air purity classes for total oil. It does not provide levels for gaseous or microbiological contaminants;

however, it does include test methods for determining the levels of a number of them should they require monitoring.

- ISO 8573 Part 1: Contaminants and purity classes.
- ISO 8573 Part 2: Test methods for oil aerosol content.
- ISO 8573 Part 3: Test methods for measurement of humidity.
- ISO 8573 Part 4: Test methods for solid particle content.
- SO 8573 Part 5: Test methods for oil vapor and organic solvent content.
- ISO 8573 Part 6: Test methods for gaseous contaminant content.
- ISO 8573 Part 7: Test methods for viable microbiological contaminant content.
- ISO 8573 Part 8: Test methods for solid particle content by mass concentration.
- ISO 8573 Part 9: Test methods for liquid water content.

The purity levels that are supplied by each gas supplier should be obtained from their websites or by contacting their sales office, the following are examples of the variation in the different purity levels commonly encountered.

- **Medicinal Air** is supplied with specifications that allow maximum levels of contaminants as set down in the European Pharmacopoeia, levels of common contaminants that are acceptable are:

  - Carbon dioxide   500 ppm V/V;
  - Carbon monoxide   5 ppm V/V;
  - Oil   0.1 mg/m³ (total hydrocarbons);
  - Nitrogen dioxide   1 ppm V/V;
  - Fluorides   1 mg/m³;
  - Sulfur dioxide   1 ppm V/V;
  - Moisture   67 ppm V/V from cylinder supply or 870 ppm from a locally supplied compressor plant using refrigeration driers (NB: refrigeration driers may not provide a sufficiently low moisture content for instrument quality air);
  - Argon   0.9%;
  - Oxygen   21% ± 1%;
  - Nitrogen   79%.

- **Instrument Grade Air**

  - Hydrocarbons   Not referenced;
  - Moisture   Less than 25 ppm;
  - Carbon Dioxide   300 ppm;
  - Methane   <5 ppm V/V;
  - Argon   0.9%;
  - Oxygen   21%;
  - Nitrogen   78%.

- **Zero Grade or Synthetic Air supplied in high-pressure cylinders manufactured by combining oxygen and nitrogen**

  - Hydrocarbons as methane   Less than 1 ppm or 0.1 ppm;
  - Moisture   Less than 5 ppm;
  - Argon   0%;
  - Oxygen   20.5% ±2%;
  - Nitrogen   Balance.

- **Zero Grade Air supplied from air compressors**

  - Hydrocarbons   Less than 0.05 ppm;
  - Moisture   Information not provided by some manufacturers;
  - Oxides of Nitrogen   Information not provided by some manufacturers;
  - Oxides of Sulphur   Information not provided by some manufacturers;
  - Argon   Information not provided by some manufacturers;
  - Oxygen   21% ± 1%;
  - Nitrogen   Balance.

Zero air specification for locally supplied sources should be determined according to the laboratory requirements. This air may contain gaseous inclusions from atmospheric sources.

Laboratory gases have far more stringent requirements than medical grade gases and specific equipment requirements are necessary to meet these specifications.

2.2.3   COMPRESSOR TYPES USED IN INSTRUMENT AIR APPLICATIONS

Before the design can proceed, it is necessary to establish what the different types of the compressor are and what the advantages and disadvantages of each are and what is available in the local or international markets.

Laboratory grade air compressors should only be selected from one type, i.e., oil-free. This encompasses oil-less compressors and oil-free compressors.

Oil-less compressors may be described as compressors that do not include a sump containing lubricating oils. These include some reciprocating units that use sealed bearings with dry sumps and generally supply low to high flow capacity compressors and include scroll, screw, water injected screw and centrifugal models.

Oil-free compressors use oil sumps and include sealed spaces between the compression chamber and the sump to prevent oil carry over into the compression chamber and are classified, however, these are not in common use.

The compressors used in the manufacture of instrument air applications are similar to those used for the production of medical air and are usually in the oil-less range. Laboratory applications require pressure and cleanliness levels not commonly found in medical applications and to meet these demands the air will need additional treatment after initial compression.

There is a common misconception that medical grade gases are particularly clean when compared to other gases. This is true when compared to gases purchased from the local gas supplier who provides industrial purity gases. Medical applications require a much higher level of cleanliness than those gases; however, laboratory grade gases demand even higher levels of purity.

The purity level of medical grade gas is 99.5%; laboratory grade gases are either HP at 99.99% or UHP at 99.9995% and have limitations of the impurities that may be considered acceptable in medical grade gases.

2.2.4   INSTRUMENT AIR COMPRESSOR PLANT DESIGN

To provide the level of purity required for instrument air, it is necessary to compress the atmospheric air, dehydrate, and then filter it to meet the equivalent levels provided in high-pressure gas cylinders. Each of these procedures requires dedicated equipment including a combination of a number of specialized items of apparatus.

2.2.5 COMPRESSOR

The various manufacturers provide similar styles and types of compressor units each of which has characteristics that meet a certain supply requirement. These vary widely and can be broken into specific areas. The driving forces that will promote the purchase of one type over another include type, pressure, price, longevity, locality, and atmospheric conditions, local support, system design, flow requirements, and instrument air quality.

2.2.6 COMPRESSOR TYPES

2.2.6.1 RECIPROCATING COMPRESSORS

Reciprocating compressors are relatively uncommon in laboratory use; there may be a number of older plants, however, these compressors are being replaced by more modern technology. Reciprocating compressors have issues with mechanical vibration, excessive noise, and heat rejection; those that are water cooled require additional services such as water treatment and associated concerns of humidity increase in plant rooms.

2.2.6.2 OIL FREE SCREW COMPRESSORS

Usually, oil free screw compressors are only used on medium to high flow applications and the smallest may be too large for most instrument air plants due to low flow requirements. Large facilities require greater volumes of instrument air and screw compressors may be the most viable alternative; however, the use of multiple smaller lower flow types are worth investigation. These compressors use screw technology with non-lubricated air ends to provide instrument air. They are available in a range of sizes that may vary between manufacturers. They are usually supplied in sound insulated canopies and are vibration free. They are provided as self-contained units that are available with built-in refrigeration driers. However, a desiccant dryer would be necessary to reduce moisture levels to the required level. They take up minimal floor space; however, ventilation is necessary for the removal of the waste heat generated during the compression cycle (Figure 2.1).

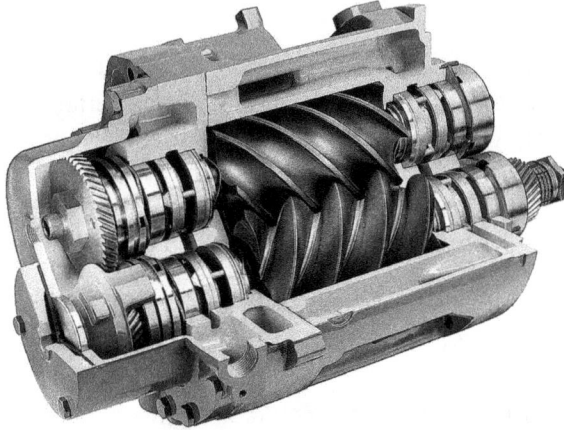

FIGURE 2.1   Non-lubricated screw compressor air end sectional view.
(Source: Atlas Copco Australasia. With permission.)

When the system demand warrants compressors of this size, they are the best alternative from a purchase value aspect and the ongoing cost for maintenance. A number of compressor manufacturers provide machines in this range. To meet the highly variable flow demands of most laboratories these compressors are available with VSD controls that may reduce operating costs during periods of low demand.

2.2.6.3 WATER INJECTED SCREW COMPRESSORS

These compressors use water as the sealing medium during the compression cycle. This allows a cooler running temperature in the compression chamber, and the condensate collected during the cooling cycle may provide some of the makeup water used in the compression chamber while in operation.

There has been some concern about the ability of the units to provide initial filtration and water treatment during operation. The manufacturer should be contacted for confirmation that this is not a concern when using their equipment. If it is the designer's intention to specify this type of compressor, then the manufacturer should be requested to provide confirmation that there are no microbiological contamination problems likely to be encountered when using their equipment. This information and written guarantees should be sought from the manufacturer before this should be selected.

2.2.6.4   SCROLL COMPRESSORS

This type of compressor is in regular use for instrument air. The units meet the demand requirements of the majority of small to medium size laboratories and the use of multiple units could be considered for large systems. They are quiet and vibration free. They are provided as self-contained units that are available with built-in refrigeration driers which may not reduce the moisture content of the instrument air sufficiently and a desiccant drier will be required. They take up minimal floor space; however, they need a continuous supply of ventilation in the plant room. Each compressor module is available with multiple elements with sophisticated control systems that provide a variable air supply by selecting sufficient components to meet the system demand (Figure 2.2).

FIGURE 2.2   Non-lubricated scroll compressor air end sectional view.
(**Source:** Atlas Copco Australasia. With permission.)

2.2.6.5   HOOK AND CLAW COMPRESSORS

A range of oil-free compressors that use a hook and claw style compression chamber are available. They are common in medical and instrument air applications that have medium to high air demands. They are relatively quiet and vibration free, provided as self-contained units that are available in package units. They take up minimal floor space but need ventilation air in the plant room. They are available in a variable speed format and may be easily interfaced with any BMS (Figure 2.3).

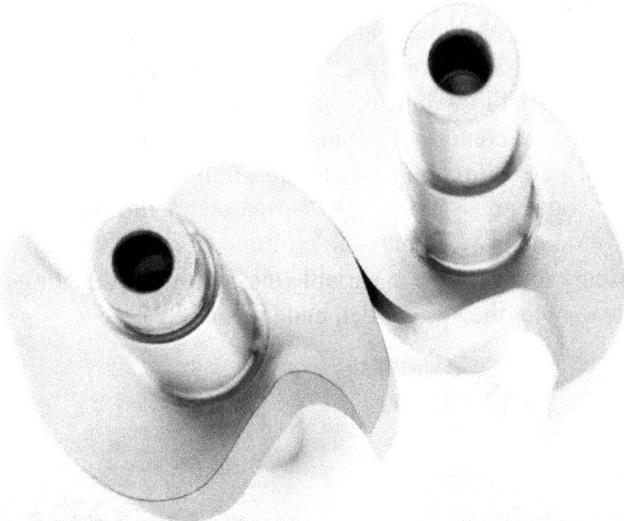

FIGURE 2.3   Non-lubricated hook and claw compressor air end.
(**Source:** Atlas Copco Australasia. With permission.)

2.2.6.6   CENTRIFUGAL COMPRESSORS

These units are preferred for use in constant demand situations not usually encountered in laboratory use. They are best suited for high flow applications where demand is constant with minimal fluctuation. They produce a constant flow under normal operating conditions. During operation the excess unused instrument air is continuously vented when it exceeds demand. They are not ideal for laboratory systems where demand is highly variable.

2.2.7   SYSTEM DESIGN PARAMETERS

2.2.7.1   PRESSURE

The pressure requirements for any compressor plant is of primary importance. However, in the majority of laboratories, the instrument air pipeline pressure is regulated to 410 kPa, which is well within the operating pressure of most off the shelf oil-free compressors. The majority of these compressors have maximum operating pressures between 700 and 1,100 kPaG. The air leaving the compressor will generally be saturated and at a temperature >200°C unless first stage cooling has been included in the compressor module in the form of an after cooler. It is also common practice to provide package compressors with built-in refrigeration driers; however, the pressure dew point from this type of drier (5°C) may be insufficient for laboratory use without the use of desiccant dryers.

The design of instrument air plants generally has a storage volume of instrument air in the air receiver. This may be the only system storage available for the system to draw upon while in operation. This is caused by the pressure regulation of the system to the pipeline pressure which would normally be regulated to a constant 410 kPa. This is usually situated in the plant room with further adjustment available at the terminal outlet if required. Higher system pressures are easily provided where required through adjustment at the pressure regulation station.

The designer will need to determine the volume of the air receiver based on the control system being used. If a VSD is specified, the volume of the receiver will be minimal; however, if the compressors are to operate on a stop/start arrangement, then the calculation is somewhat more involved. The calculation needs to take into consideration are:

- Cut-in pressure and cut-out pressure and the volume of instrument air required from the compressors during the operating cycle. If the pressures are close together the compressors will short cycle so the system design needs to factor the cycle period to meet the manufacturer's maximum number of cycles per hour or as per their recommendations.
- System operating pressure.
- Approximate volume of the pipeline attached to the vessel prior to the system pressure regulator is provided.

- If multiple compressors are used in the design, they should be operated on a cyclic basis to ensure even wear of the units. The calculation of the number of starts per hour should take this into consideration.

2.2.7.2   PRICE

The price of any compressor is always a major concern. However, if the unit is being installed in harsh climatic conditions, it is recommended to select the best quality equipment using robust construction to provide protection from adverse conditions, better longevity, and serviceability for the future. Unfortunately, in many instances, budgetary constraints will override this with future cost blowouts for repairs and maintenance plus the downtime of vital equipment and the associated disruption that accompanies it.

2.2.7.3   LONGEVITY

The lifetime expectancy of a compressor will depend on a number of factors; quality of the unit, locality of the installation, local atmospheric conditions, maintenance provided and its availability, expected workload, and running time. The better the quality the longer the unit will stand up to the rigors of hard use. The opposite to that is to supply the best quality compressor in a situation where it will operate minimally under good operating conditions with regular maintenance and service. This may be considered unnecessary as a reasonable quality compressor will adequately provide this service. The choice of equipment available in the local marketplace may reduce the options available for this alternative.

2.2.7.4   LOCALITY AND ATMOSPHERIC CONDITIONS

The site conditions will vary considerably from installation to installation. For example, a plant being installed in tropical locations in an open plant room will be subjected to extremes of humidity, atmospheric contaminants, and temperature that an equivalent compressor in a cool climate would not expect to encounter.

Installation in an air-conditioned plant room may be necessary if the compressor plant and equipment selected has maximum operating temperature limitations. Water-cooling plant is also an option that will need special consideration as water treatment will be necessary as well as the additional costs and safety concerns that go with it. Any location that has high humidity and temperature will also adversely affect the efficiency of the driers being used and allowances must be made at design time to accommodate these.

2.2.7.5 LOCAL SUPPORT

The ability of local service providers to maintain and service the plant will play a significant role in the selection of the manufacturer. Spare parts that need to be purchased from warehouses that are distant or overseas from the installation site is unacceptable; especially, if the installation is for research use, the staff relies on the ongoing supply being readily available. Extended shutdowns while waiting for repairs and spare parts should be avoided at all costs. In some instances, it may be advisable to provide multiple compressors and associated equipment to prevent this from occurring.

2.2.7.6 FLOW REQUIREMENTS

The system designer must take all of the above into consideration when specifying any equipment. Flow rates are difficult to calculate as usually there will be no empirical data to call on plus little in the way of documented references available from the laboratory. System flows can only be based on the equipment that has been selected for use in the laboratory and it is essential that the suppliers be contacted to provide this essential data. The equipment used in any laboratory is available in a range of capacities and as such, there is little similarity between specific models of the same type of equipment not only from the various suppliers as the same supplier will invariably provide a range of equipment.

2.2.8 INSTRUMENT AIR TREATMENT

After leaving the compressor and air receiver, the air would normally be saturated and approximately 10°C warmer than the ambient conditions. It would still contain contaminants found at the air intake including moisture, particulates (e.g., dust), gases (e.g., CO), and microbes that have managed to pass through the high temperatures of compression and other contaminants that could be included in the local air supply. To remove these contaminants the instrument air will require dehydration and filtration.

2.2.8.1 MOISTURE REMOVAL

When designing an instrument air plant, the selection of the compressor type for the installation poses additional variables. The considerations such as noise and vibration transmission to the building, heat rejection, power availability and floor space are just some that will need to be contemplated. In most instances, it will be necessary to install desiccant drying for the instrument air to provide suitably dry air. When these dryers are used, there will be a need to increase the plant design flow rate to compensate for the additional consumption of the instrument air used in the drying process. The treated air downstream of the desiccant dryer is used to recycle the desiccant beds used in the drying process. This is usually 18% of the design flow rate of the drier for each drier that is connected to the system. The volume of air used during the recycling process is constant and not relative to the flow of air passing through the drier at any time (Figure 2.4).

Moisture is a variable that has two acceptable criteria; instrument air from high-pressure cylinders has an approximate moisture content below 67 ppm (the equivalent of an atmospheric dew point of minus 48 degrees C). Air from a compressor plant using a refrigeration drier has an approximate pressure dew point of 5°C at 800 kPa pressure (equal to 870 ppm). This is two completely different levels of contamination, 67 ppm versus 870 ppm; both of which pass the requirements of many national and international standards for medical air. However, the 870 ppm is not sufficiently dry for laboratory applications where purity of the air supply is of significant importance.

The level of moisture vapor will directly affect the air stream and in a laboratory. It may impact the accuracy of the equipment using the instrument air, if it is sensitive to minor volumes that can be measured in ppm in the air stream.

A refrigerant dryer uses a standard refrigeration system to cool the air stream. This system cools the instrument air to a temperature of between 4 and 5°C causing any moisture to condense to this temperature, under pressure ready for removal from the pipeline. This temperature range must not fall below zero or freezing of the condensate will allow ice to form and blockages will occur in the pipeline. Refrigeration dryers are not able to remove the moisture to similar levels to that provided by desiccant dryers. If a moisture content of approximately 870 ppm is acceptable, this method would be acceptable (this ppm level is dependent on the system pressure and will vary with pressure fluctuation).

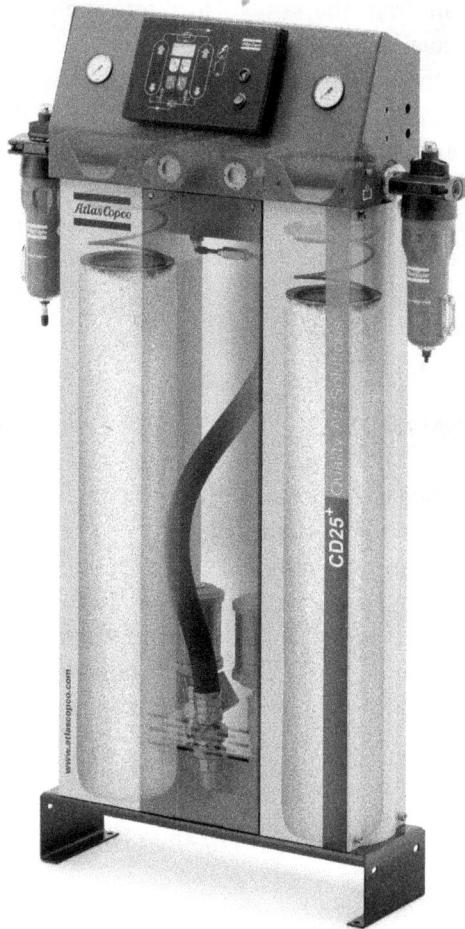

FIGURE 2.4   Desiccant dryer.

(**Source:** Atlas Copco Australasia. With permission.)

Desiccant dryers are commonly used to remove this moisture in instrument air systems. These dryers are available in temperature ranges that will provide suitable levels of residual moisture. The standard range used for laboratory service has a dewpoint setting of −70°C. This type of dryer uses dual desiccant beds and operates on a recycling process; using one bed for drying the saturated air while it uses a portion of the already dried air stream to dehydrate the alternative desiccant bed during operation. The desiccant beds are cycled between the incoming air stream and the previously dried air from the system to purge the desiccant. The more recent models are available with an air dew point monitoring that reduces the drying airflow to suit the demand and the drier will shut down when the demand for air ceases. Each dryer of this type will use approximately 18% of the free air volume it produces during its recycling process to dehydrate the offline desiccant bed. This flow rate will need to be taken into consideration when calculating the capacity of the compressor plant. If multiple desiccant dryers are installed and operate concurrently, the air supply used for the recycling process will be the combined total of 18% of each dryer's design flow rate.

2.2.8.2   CARBON DIOXIDE AND CARBON MONOXIDE

The removal of carbon dioxide and carbon monoxide is an option that can be provided as part of the desiccant dryer. These driers are necessary if the design intent requires the removal of carbon monoxide. The catalytic converter used to convert the carbon monoxide to carbon dioxide may not operate in saturated air systems that have the moisture levels supplied by refrigerant drier types. It is necessary to contact the manufacturer to confirm the suitability of the filters proposed and the conditions under which they operate.

2.2.8.3   HYDROCARBON CONTAMINATION

The selection of the plant to be provided is of critical importance. The levels of hydrocarbons should be nil as laboratory grade air compressors should be oil-free or oil-less compressors.

During the commissioning period, an oil level of below 0.5 mg/m^3 could be a pass mark for the plant based on the European Pharmacopoeia

[29] for medicinal air supplied from high-pressure cylinders and is what may be found in instrument air in high-pressure cylinders.

If hydrocarbon contamination in the instrument air supply is suspected, it may be removed at the terminal outlet using molecular sieves that can be fitted into the pipeline. If hydrocarbon contamination is left in the air stream, it may render a laboratory instrument air supply unusable. The use of molecular sieves or activated carbon filters may be an option at the terminal outlets. However, these filters may only be available with very low flow rates and only suitable for connection to single equipment use.

Commissioning test results from a compressor supply should confirm the oil content as zero. Any hydrocarbon concentration would indicate a plant fault or an air intake problem that would need to be resolved prior to commissioning.

To prevent airborne gaseous contamination, it is recommended that an independent fresh air supply is provided from an exterior source; usually above the building or at a high level away from any local sources of contamination such as roads or any nearby industrial plant or manufacturing facility.

2.2.8.4   PARTICULATE REMOVAL

Entrained particles and vapors can be removed by using standard filters available from the majority of the instrument air filter suppliers. When operating under normal circumstances this is carried out in three stages: pre filtration, activated carbon filtration, and final filtration.

2.2.8.4.1   Pre Filtration

Pre-filters can be installed either prior to or after the refrigerated dryers to remove particulates above 5-micron size. This protects the systems condensate removal system from becoming fouled with particles that have passed through the compressor air intake (which would have a nominally 20-micron filter). The filters are provided with a differential pressure gauge to monitor the filter element's efficiency and an integrated automatic drain to remove moisture during normal operating conditions.

2.2.8.4.2   Activated Carbon Filtration

Activated carbon filters use a molecular sieve to trap odors and oil vapors that may have entered the pipeline at the compressor intake. They do not include a differential pressure gauge, as the filter element is effectively a molecular sieve and its life is determined by the volume of air likely to pass through the unit during a predetermined time. These filters do not lose pressure due to the condition of the filter element. An automatic drain should not be necessary, as any moisture will be removed by the desiccant dryer prior to the air entering the filter.

2.2.8.4.3   Final Filtration

Final Filters are installed after all other dehydration and filtration have been carried out. They remove particulates down to 0.01 microns with an efficiency of 99.9995%. This filtration size is capable of removing almost all known bacteria and some viruses.

Note: When removing and or replacing these filter elements it should be standard procedure to wear protective clothing suitable for use with biohazardous materials.

2.2.8.5   MONITORING OF THE AIR STREAM

In some installations, it will be necessary to monitor the air stream for purity; monitors that measure the particulates, dew point, carbon monoxide, and carbon dioxide levels are readily available. This may be required if the air supply is of such a nature as to include high levels of these contaminants from the local atmospheric air. Sources that may be located near to large motorways, local industrial manufacturing facilities, and storage facilities for various volatile chemicals need to be investigated during the design stage. And, local atmospheric tests for particulates, oil, and vapors should be recommended prior to the commencement of the system design and definitely before installation proceeds.

There are commercially available monitoring systems that will measure most contaminants and provide digital outputs and displays to give a continuous readout of the health of the laboratory instrument air supply.

2.2.9   SYSTEM DESIGN FLOW RATE

The design flow rate for the instrument air supply will affect the final selection of the type of plant that will be the most efficient system to meet the laboratory's demand. The options that are available are limited to basically two alternatives. They are a high-pressure supply system or a compressor plant, each of which has advantages.

2.2.9.1   HIGH-PRESSURE INSTRUMENT AIR CYLINDER SUPPLY

If ease of installation and minimal construction costs are a major concern, then the simplest system is a high-pressure manifold designed for use with the gas purity level selected. The number of cylinders will be determined by the availability of supply and the provision of storage space for the manifold.

It is recommended that automatic changeover high-pressure manifolds be selected for use with instrument air supplies to provide a guarantee of supply when the system is in use.

The number of cylinders required will be determined by the expected flow rate and should allow for the maximum delivery time turnaround for a replacement bank of cylinders. Cylinders are available from most gas suppliers as individual cylinders or as packs with cylinders fixed into a frame and pre-piped to a single connection valve. This type of layout will require access for a forklift or similar transport system.

2.2.9.2   LOW-PRESSURE COMPRESSOR SUPPLY

Should the system require substantial flows such as would be encountered in a major university or research facility? Then, a compressor plant is the best choice! It may have initial cost implications; however, the alternative supply costs for significant numbers of gas cylinders over the life of the facility would indicate that this would be the preferred option.

The design of the plant room for an instrument air compressor plant will require additional services that would not be required for a standard industrial installation. The plant room itself should be a separate independent room to ensure there is no local contamination from adjacent plant servicing the facility.

An independent fresh air supply should be provided for the room to provide sufficient air changes to remove the waste air used by the compressors and plant during operation. The compressor manufacturer must be consulted to determine the volumes of air required for this purpose.

A dedicated compressor fresh air intake may be necessary depending on the location of the plant room. Local ambient conditions may be contaminated with carbon monoxide, noxious fumes, or high concentrations of contaminants. These impurities would contaminate the instrument air supply sufficiently to create a variation in the makeup of the air supply and corrupt any results provided by the equipment using the gases.

The plant should be connected to an uninterruptable power supply (UPS). The supply must be maintained when in use to protect the equipment being used and as a safety measure to ensure that mixed gases such as is used in an AAS, (where the mixed acetylene and instrument air or nitrous oxide are used) which if disrupted could allow undiluted acetylene to be released into the laboratory.

It is possible to monitor the instrument air supply for hydrocarbons, moisture, and atmospheric contaminants in the plant room with remote alarms at points of use.

All laboratories should have gas sensors fitted for all gases used in every location. These should include sensors for low oxygen, all flammable gases, all toxic gases, and specific sensors for carbon dioxide and nitrous oxide [7–13].

2.2.9.3   COMPRESSOR PLANT LAYOUT AND CONFIGURATION

The selection of the plant and system design will be dependent on the client. Many instrument air plants are designed on the layouts used for medical air compressor plants which may not always be appropriate. A medical air plant is a life support system and is designed with at least 100% redundancy for all items of equipment. This may not be a necessity for instrument air supplies. It may be possible to select from a number of configurations to suit the project.

Examples of the alternatives include:

- A single compressor and associated items of equipment with no backup support designed to provide 100% of the system design flow rate (Figure 2.5).

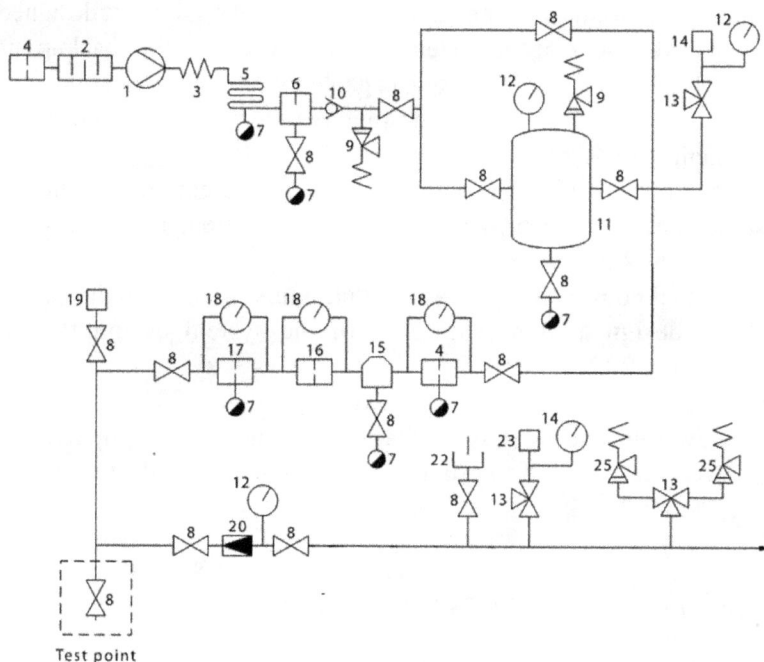

1 = Compressor

2 = Inlet silencer

3 = Vibration isolator

4 = Pre-filter 5.0 micron

5 = Integrated after-cooler

6 = Moisture separator (optional)

7 = Automatic condensate drain

8 = Equipment isolation valve

9 = Primary pressure safety valve

10 = Non-return valve (optional)

11 = Air receiver

12 = Pressure gauge

13 = Pressure switch valve

14 = Pressure switch low alarm

15 = Desiccant drier

16 = Activated carbon filter

17 = Final-filter 0.01 micron

18 = Differential pressure gauge

19 = Dew point sensor

20 = Pipeline pressure regulator

21 = Pipeline isolation valve

22 = Service facility connection

23 = Pressure switch failure alarm

24 = Pipeline pressure gauge

25 = Pipeline pressure safety valve

NB: The use of NRV's (item 10) downstream of locally controlled package compressors is not recommended.

FIGURE 2.5   Single compressor plant schematic.

(AS 2896–2011 Figure 2.7. A modified. © Standards Australia Limited. Copied by James (Jim) Moody with the permission of Standards Australia under License 1703-c120.)

- A single compressor and associated items of equipment designed to provide 100% of the system design flow rate with a backup high-pressure cylinder manifold capable of providing instrument air supply to meet the minimum time requirement to provide replacement cylinders (Figure 2.8).
- A dual compressor and associated items of equipment with each set designed to provide 100% of the system design flow rate (Figure 2.6).
- A dual compressor and associated items of equipment with each set designed to provide 50% of the system design flow rate (Figure 2.6).
- A triplex compressor with a capacity to meet 33% of the design flow rate and dual sets of associated items of equipment with each set designed to provide 50% of the system design flow rate (Figure 2.7).

2.2.9.4   *TRIPLEX COMPRESSOR PLANT LAYOUT*

This type of configuration would include isolation valves on either side of each item of equipment. This allows the removal of any item of the plant from service for maintenance and repairs without disrupting the instrument air supply (Figure 2.9).

2.3   LABORATORY VACUUM PLANT

2.3.1   *DESCRIPTION OF LABORATORY VACUUM*

Vacuum systems used in laboratories have a variety of uses. The majority are for general suction that in times past was generated by venturi operated devices connected to water taps and used as and when necessary. These generated a limited vacuum level and consumed considerable volumes of water (Figure 2.10).

Current vacuum constraints are more challenging requiring higher levels of vacuum and are often specific to meet the stipulations of particular equipment. The design of any vacuum system must perform according to the laboratory's requirements. This will need to encompass a range of

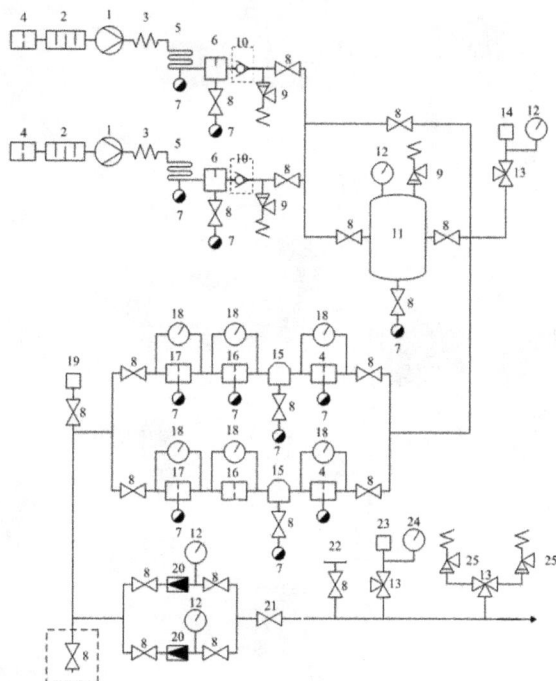

1 = Compressor
2 = Inlet silencer
3 = Vibration isolator
4 = Pre-filter 5.0 micron
5 = Integrated after-cooler
6 = Moisture separator (optional)
7 = Automatic condensate drain
8 = Equipment isolation valve
9 = Primary pressure safety valve
10 = Non-return valve (optional)
11 = Air receiver
12 = Pressure gauge
13 = Pressure switch valve

14 = Pressure switch low alarm
15 = Desiccant drier
16 = Activated carbon filter
17 = Final-filter 0.01 micron
18 = Differential pressure gauge
19 = Dew point sensor
20 = Pipeline pressure regulator
21 = Pipeline isolation valve
22 = Service facility connection
23 = Pressure switch failure alarm
24 = Pipeline pressure gauge
25 = Pipeline pressure safety valve

NB: The use of NRV's (item 10) downstream of locally controlled package compressors is not recommended.

FIGURE 2.6 Dual compressor plant schematic.

(AS 2896–2011 Figure 2.7A. © Standards Australia Limited. Copied by James (Jim) Moody with the permission of Standards Australia under License 1703-c120.)

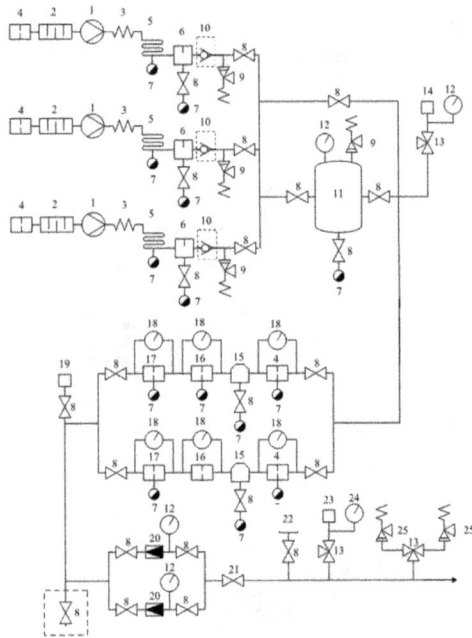

1 = Compressor
2 = Inlet silencer
3 = Vibration isolator
4 = Pre-filter 5.0 micron
5 = Integrated after-cooler
6 = Moisture separator (optional)
7 = Automatic condensate drain
8 = Equipment isolation valve
9 = Primary pressure safety valve
10 = Non-return valve (optional)
11 = Air receiver
12 = Pressure gauge
13 = Pressure switch valve

14 = Pressure switch low alarm
15 = Desiccant drier
16 = Activated carbon filter
17 = Final-filter 0.01 micron
18 = Differential pressure gauge
19 = Dew point sensor
20 = Pipeline pressure regulator
21 = Pipeline isolation valve
22 = Service facility connection
23 = Pressure switch Failure alarm
24 = Pipeline pressure gauge
25 = Pipeline pressure safety valve

NB: The use of NRV's (item 10) downstream of locally controlled package compressors is not recommended.

FIGURE 2.7   Triple compressor plant schematic.

(AS 2896–2011 Figure 2.7B. © Standards Australia Limited. Copied by James (Jim) Moody with the permission of Standards Australia under License 1703-c120.)

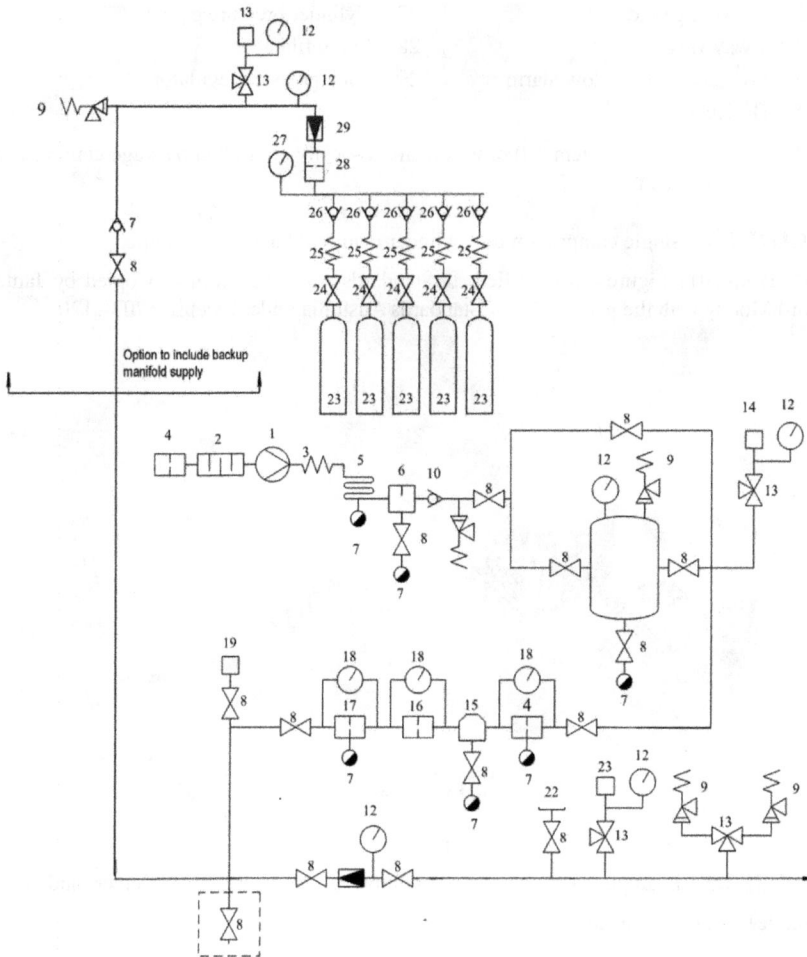

1 = Compressor
2 = Compressor inlet silencer
3 = Flexible connector
4 = Pre-filter
5 = Instrument air after-cooler
6 = Moisture separator
7 = Automatic drain
8 = Isolation valve
9 = Line pressure relief valve
10 = Non return valve (optional)
11= Air receiver

16 = Activated carbon filter
17 = Final filter
18 = Differential pressure gauge
19 = Dew point sensor
20 = Differential pressure gauge
21 = Pressure regulator
22 = Service inlet facility
23 = Pressure switch: instrument air failure
24 = Cylinder isolation valve
25 = Flexible lead
26 = High pressure non return valve

12 = Pressure gauge   27 = Cylinder pressure gauge
13 = 3 way valve   28 = Inline filter
14 = Pressure switch – low alarm   29 = High pressure regulator
15 = Desiccant drier

NB: The use of NRV's (item 10) downstream of locally controlled package compressors is not recommended.

FIGURE 2.8   Single compressor and cylinder manifold backup schematic.

(AS 2896–2011 Figure 2.6 modified. © Standards Australia Limited. Copied by James (Jim) Moody with the permission of Standards Australia under License 1703-c120.)

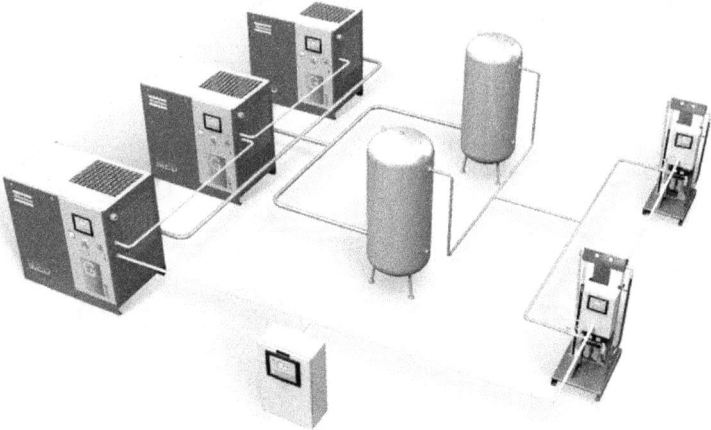

FIGURE 2.9   Example layout for a triplex compressor plant with dual receivers and driers.
(**Source:** Atlas Copco Australasia. With permission.)

pressure and flow specifications to meet the demands of individual items of equipment.

The vacuum pipeline may be connected to a number of laboratories each of which will have demands on the system that will cause system pressure variation during use. The variation in system pressure will impact on the various laboratories that may be located on different floors and in locations spread throughout the building. It is necessary to ensure that internal pipeline pressure variation does not cause backflow between the various branches of any multi-connection system. Backflow can be caused by locally undersized pipelines or plant or by sudden demand in single locations that may cause the system pressure to vary widely throughout the facility. This localized usage may create a demand that reduces the local pressure and causes a backflow into an adjacent laboratory or between that location and other areas with lower system demand. This can be overcome using local pressure positive non-return valves that prevent backflow between laboratories or between individual taps or valves or branches in the event of a pressure fluctuation. These valves may also require some form of filtration to remove particulates that may be drawn into the system from preventing the non-return valves from sealing.

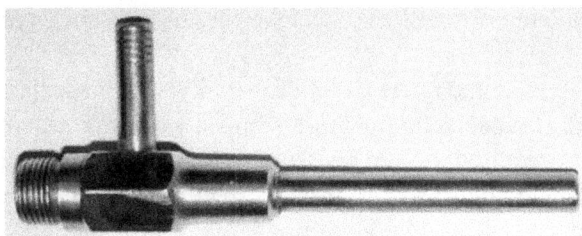

FIGURE 2.10   Water operated venturi suction device.

As the vacuum system design needs to suit the application, the rules that can be applied for calculation of flows and diversity factors are variable. The pressure, location, numbers, and types of terminal units need to be selected according to the expected demand. In medical applications, the vacuum system is designed to provide a pressure of –60 kPaG with a flow of 40 liters per minute per terminal in the majority of locations and this would be adequate for a general purpose laboratory. If the laboratory is a teaching classroom and a number of students would be working on similar procedures concurrently, the diversity would be very high; probably around 90%. If the laboratory is a research laboratory with multiple benches and with a limited number of staffs, then the diversity would be considerably lower between 20% and 30%.

The operating pressure range of –60 kPa to –85 kPa (or –70 kPa if operating with a variable speed drive – VSD) may be suitable for general purpose laboratory vacuum system practice. However, it will not provide the levels of vacuum required to evaporate some of the volatile organic compounds (VOCs) used in some procedures nor meet the demands that may be applied when a class of students using the same equipment use the laboratory vacuum system simultaneously. It is absolutely necessary to determine what equipment is intended to be used; the numbers and types of users and design the system accordingly.

The general-purpose type of vacuum system could be of a similar system design to that used in medical applications. However, there are other considerations that must be explored and taken into consideration and the correct pumps selected for that purpose.

For pressures from zero to −85 kPa, it is good practice to select a dry running vane or dry running claw pump designed to operate 24/7 using a variable speed drive. It is essential to select the correct type pump as some models may have wear issues on the ends of the vanes that can create major inefficiencies resulting in flow and pressure losses over time.

For pressures in the range of −85 to −95 kPa, an oil sealed type pump would be the ideal selection using a stop-start control system, pre-set at the minimum required system pressure. If the pumps are to be used for corrosive service, the selection of the correct material is essential and an oil seal model would not be appropriate and a dry running screw pump manufactured from compatible materials may be the best option.

Note: The use of VSD controls with lubricated vacuum pumps is not recommended. Lubricated pumps require minimum RPM to provide the centrifugal force necessary to maintain the lubricant seal in the outer ring of the pump. Should the pump reduce the RPM it will not provide sufficient centrifugal force to maintain the seal and the vacuum may fail.

The minimum operating pressure required will be the determining factor for pump selection as will the use of the system.

As an example, the systems used to evaporate aggressive solvents may react with the materials that the pump is manufactured from. This should be taken into consideration when selecting the pump type. In these situations, the pump will need to be manufactured from compatible materials, e.g., stainless steel, with appropriate working pressures. It is essential that the proposed service be discussed with the manufacturers of any equipment that is being specified in the design brief.

The following variables provide some of the restrictions that may assist in the correct vacuum pump selection for an installation based on the demands of the system from a pressure, flow, and compatibility perspective. The final selection will be up to the designer and a poor choice at the design stage could render the vacuum system expensive to operate and ineffective.

1. The first variable is the operating pressure required by the laboratory, this will determine the pump type to be used.
2. The second variable is the design flow demand for the system, this is dependent on the client's requirements; and without this information, it is not possible to select the correct plant. The various vacuum levels are separated into two main areas, rough vacuum, and high vacuum. Details on these levels and their respective pressure ranges are discussed later in this chapter.
3. The third variable is what materials the pump will need to extract, e.g., air, air containing vaporized liquids, air water mixes or vaporized aggressive chemicals. For particularly aggressive contaminants the pumps and the pipeline materials need to be selected to counterbalance any corrosive action. It is highly recommended that consultation with the laboratory staff is undertaken to determine what they will be using the vacuum system and any reagents that are likely to be encountered.

2.3.2   VACUUM PUMP SELECTION

When selecting a pump it is not advisable to select one that has a minimum vacuum operating pressure on the upper or lower limit of the system design. A pump that can operate to a minimum pressure of –99.5 kPa should not be selected for continuous use at that pressure.

For each type of vacuum pump available, the major differences are the level of vacuum they can achieve and the method; they use to evacuate the system. In general, duty dry running vane type pumps operate to approximately –80 kPa, claw type dry running pumps are capable of a range between –80 and –94 kPa depending on pump size. Oil flooded vane pumps operate to –99.5 kPa and screw vacuum pumps operate to 0.01 hPa. These types of pumps cover most laboratory vacuum systems adequately; however, if corrosive chemicals are likely to be encountered, it would be advisable to select a pump that has the option of providing protective plating or coatings. Some pump types are able to remove limited volumes of fluids that may inadvertently find their way into the pump body; however, that is not recommended. It is also likely that it is not possible to maintain a constant temperature in the exhaust pipeline due to cooling that is experienced downstream of the pumps and condensation in the exhaust air stream will occur and allowance must be made for its removal. The installation of liquid traps and particulate filtration in the pipeline in any location where this may occur is recommended to prevent contaminants from entering the pump body itself. The use of particulate filters and activated carbon filtration is recommended where odors or hydrocarbon vapors may be encountered. The removal of oil vapors is essential when operating at vacuum levels below –60 kPa. The temperature of the exhaust gas from the pump operating at high vacuum can be extreme and it is advisable to design the system with this in mind in the early stages of the project plan. Discussion with the pump supplier's engineering staff and filter manufacturers is necessary to ensure any contamination that may be carried over into the exhaust pipeline is prevented. Vacuum exhaust filtration is available and should be included, if necessary, to remove any carryover after being exhausted from the pumps. If there is a likelihood of carryover of VOCs from the pipeline, it is advisable to provide local condensers or filters in the laboratory prior to the connection to the pipeline.

All of the variables that may occur must be allowed for in the pipeline design at the outset of the project as they will impact on the selection of materials and plant to be used.

As an example of the differences between vacuum pump types, the following provides examples of the various models available. We have used the Busch range of vacuum pumps for examples.

2.3.3 TYPES OF VACUUM PUMPS

The following information is reproduced with the permission of Busch Vacuum Pumps and Systems.

2.3.3.1 BUSCH SECO DRY RUNNING VANE TYPE PUMPS

Compact, reliable, and extremely powerful – these are the stand-out features of dry-running Seco rotary vane vacuum pumps. Their lubricant-free design makes them ideal for industrial applications in which rapid and reliable vacuum services are required (Figure 2.11).

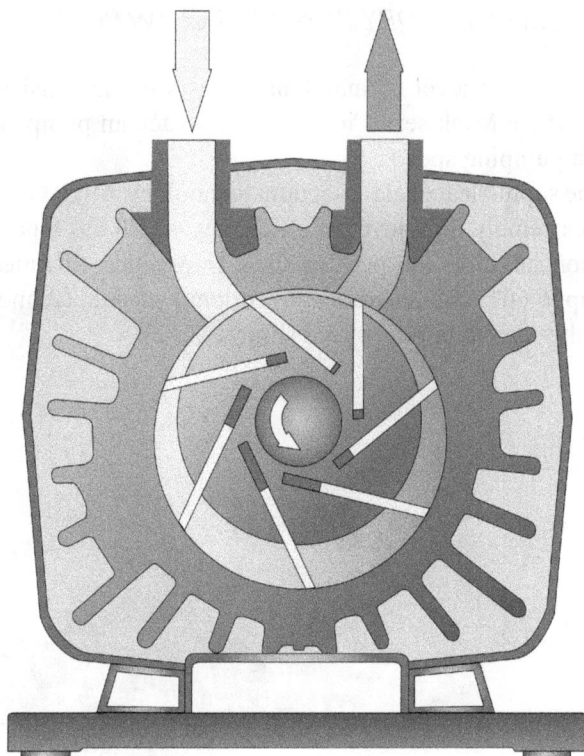

FIGURE 2.11 Busch Seco Dry running vane, cross-sectional view.
(**Source:** Busch Australia. With permission.)

Busch Seco rotary vane vacuum pumps are characterized by oil-free operation and high levels of availability and operational reliability. This is a result of design features such as hard-wearing and self-lubricating special graphite rotary vanes, robust construction and lifetime-lubricated bearings. The compact dimensions of the pump allow installation to be carried out almost anywhere with ease, whilst the energy-efficient drive ensures economical operation.

Easy to service: Maintenance is easy, and can be carried out by the operator. Apart from regular checks and the replacement of rotor vanes and filters at recommended service intervals, no additional maintenance is required. Seco rotary vane technology is also used in compressors: Seco SD rotary vane compressors generate an overpressure of 1.5 bar (g) oil-free.

2.3.3.2   BUSCH MINK DRY RUNNING CLAW PUMPS

Combine the lowest level of maintenance as well as consistent performance. The whole Mink series includes claw vacuum pumps from 40 up to 1000 m³/h pumping speed.

Due to the sophisticated claw vacuum technology, Mink vacuum pumps achieve an extremely high level of efficiency, which has a positive effect on energy consumption and performance. In practice, this means energy savings of up to 60% compared to conventional vacuum technology when operated at the same pumping speed (Figure 2.12).

FIGURE 2.12   Busch Mink dry running claw pumps cross-sectional view.
(**Source:** Busch Australia. With permission.)

An additional benefit of claw vacuum technology is the virtually maintenance-free operation due to the non-contact operating principle. None of the moving parts inside the vacuum pump come into contact with one another; meaning there is no wear at all.

The need for maintenance, such as the inspection or replacement of worn parts, is completely eliminated. Due to the completely dry compression without the need for any operating fluids in the compression chamber, there are no costs for purchase, provision or disposal. Mink claw vacuum pumps are air cooled.

The high operational reliability and long life cycles of Mink claw vacuum pumps are also a result of their non-contact compression without operating fluids. Due to the wear-free operation, vacuum, and suction performance remains consistently high throughout a life cycle of the pump. A smart silencer concept enables quiet operation.

Technical specifications: The Mink vacuum pumps use two claw-shaped rotors turning in opposite directions inside the housing. Due to the shape of these claw rotors, the air or gas is sucked in, compressed, and discharged. The claw rotors do not come into contact either with each other or with the cylinder in which they are rotating. Tight clearances between the claw rotors and the housing optimize the internal seal and guarantee a consistently high pumping speed. A synchronization gearbox ensures exact synchronization of the claw rotors. Mink vacuum pumps are driven by a directly flange-mounted asynchronous motor, with an efficiency class IE2/IE3.

Industrial vacuum generation for many applications: Mink claw vacuum pumps are available in a wide range of sizes. Special models for certain applications such as dust and gas explosion protection, high water vapor contents, gas tightness, increased oxygen contents etc. are also available.

2.3.3.3   BUSCH COBRA – DRY SCREW VACUUM PUMP

The ideal vacuum pump for many processes in the chemical industry and pharmaceutical industry as well as applications in other sectors of industry.

Description: The COBRA dry screw vacuum pumps are available in three models. The COBRA NC and AC models are specifically designed for difficult applications in the chemical and pharmaceutical industries, while the COBRA NP model is designed for a broad range of non-aggressive industries where a relatively high vacuum level is required. The NC and NP models are available as direct cooled or with closed-loop air cooling, while the AC model is direct cooled only. All COBRA models operate at any pressure from atmosphere down to ultimate pressure (Figure 2.13).

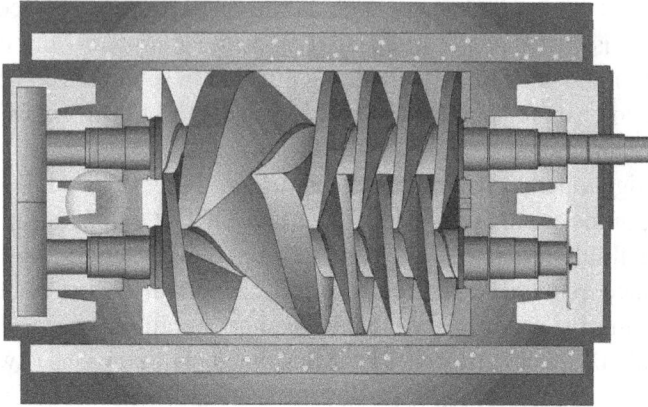

FIGURE 2.13 Busch Cobra-dry screw vacuum pump, cross-sectional view. (**Source:** Busch Australia. With permission.)

The inclusions are:

- Oil-free operation – no disposal costs, easy product recovery, environmentally friendly
- Simple screw design – fewer parts, easier maintenance
- Single stage – no intercoolers required
- Small footprint – for less space requirements
- Non contacting parts – for longer pump life
- ATEX certified for Zone 0 and Zone 1
- Oil-free and contactless sealing
- Simple and robust construction
- Optimally construable due to the many model sizes and the version available

Busch offers the largest selection of ATEX-certified, mass-produced screw vacuum pumps for zones 0 to 1.

2.3.3.4 BUSCH R5 OIL LUBRICATED ROTARY VANE VACUUM PUMPS

There are over 2.5 million R5 vacuum pumps worldwide that provide dependable service under the harshest industrial conditions (Figure 2.14).

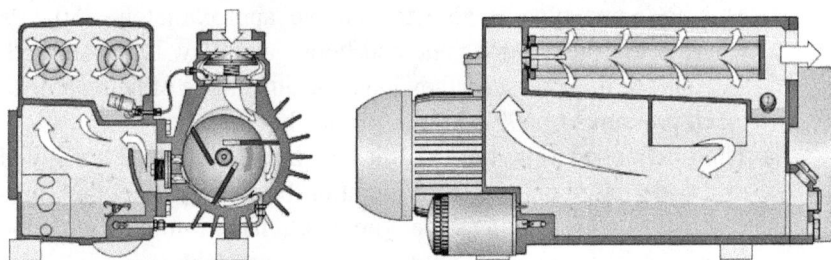

FIGURE 2.14   Busch R5 Oil lubricated rotary vane vacuum pump cross-sectional view. (**Source:** Busch Australia, with permission.)

Safe and cost-effective: Rotary vane technology has been continuously developed and optimized by Busch over the decades, with the emphasis on operational reliability and efficiency. R5 rotary vane vacuum pumps are known throughout the industry for their modern and energy-efficient vacuum generation, in a wide range of applications. Whether for intermittent or continuous use, you can rely on the R5.

Proven: Compact R5 vacuum pumps owe their robustness to proven rotary vane technology with recirculating oil lubrication. This guarantees a consistently high vacuum level which can cope with the toughest operating conditions. When fitted with a gas-ballast valve (optional), vapors can be pumped without condensation.

2.3.4   SELECTION OF VACUUM PUMP TYPE

The pumps available from the various suppliers in many instances utilize similar operating principles and are suitable for the same laboratory applications. However, each type may provide advantages where a number of pump types can meet the system design criteria.

The use of oil-flooded vane type pumps, dry vane type pumps plus hook and claw type pumps are all frequently used in laboratory service for similar operating parameters. Each pump type has the ability to provide a system pressure of −80 kPa which is the minimum pressure commonly encountered in laboratory use where there are no aggressive contaminants likely to be encountered. Each pump type has its strengths and weaknesses, and the examples are:

(i) The dry vane type is able to achieve approximately −80 kPa whereas the oil flooded vane and hooks and claw types are both suitable for this pressure, they have the advantage of being able to reach pressures of −99.5 kPa or lower.

(ii) The oil flooded vane type and the hook and claw type are suited to a wide range of stop-start operation between pre-set pressures whereas the dry running vane type would have limited ability due to the minimum operating pressure range available.

(iii) The dry running vane type and the hook and claw type using a VSD and running at a constant pressure have the advantage of providing a far better efficiency than a pump operating at lower pressures.

(iv) The water sealed and oil flooded vane type pumps are not suitable for use with VSDs.

(v) The service and maintenance of the oil flooded vane type and the claw and hook type are likely to be lower than the dry running vane type as the pump component parts do not come into direct contact with the body of the pump.

2.3.5   VACUUM PLANT CONTROL SYSTEM

The operational control system for a vacuum plant is under the direction of the design engineer, this can provide a simple stop/start controller through to VSD controlled systems with BMS overview. There are a number of inclusions that should be considered by the design engineer to ensure the correct function of the plant and provide some level of system management that would be advantageous. The following inclusions are only some of those available that are worth consideration:

- The suction plant should be operated through a central control panel mounted adjacent to the plant. The devices located on the door of the panel, and within, should be laid out logically. And in addition to the following requirements, include a mains isolator, circuit breakers for each pump control circuit and pressure switches or a transducer with adjustable range and differential settings depending on the operating management system being stop/start or variable speed drive.

- Controls should be arranged to ensure continuous operation of the plant. The failure of one pump occur it will not interfere with the correct operation of the remaining plant.

- Manual-off-auto selector switches should be provided for each pump with the manual selection overriding the pressure switch or transducer control of the respective pump. The vacuum pump should be under the control of the pressure switch or transducer when the selector is in the Auto position.
- The lead/lag/standby roles of the vacuum pumps can be rotated to ensure even running time across the plant. The lead pump will respond to the lead pressure switch or controlled through the transducer and the system control. The remaining pump or pumps are automatically operated in sequence (lag/standby duty) on demand. The control system should automatically sequence the lead pump after each cycle.
- Each pump should be provided with a "Green" run light and a "Red" fault light. The fault light should be illuminated in the event of the pump stopping for any of the following reasons:

  - Motor over-current;
  - Failure to start; and
  - Low oil level (where applicable).

- Control of the pumps should isolate any faulty unit and require that the faults will require manual intervention to clear the fault and reset the pump.
- Visual management through touch screen controls provides a visual indication of any fault condition.
- Hours run indication should be provided for each pump.
- Provisions for indication of the expiration of the manufacturers specified service interval. This indication must not interrupt the operation of the plant and is not cleared in the event of a power failure. Clearing this alarm should involve an independent procedure. A volt-free contact provides clean terminals to allow remote indication of this alarm.
- To prevent the vacuum motors from starting more than ten times per hour, a minimum run time for the pumps should be built into the pump control system after consultation with the manufacturer.

2.3.6   VACUUM PIPELINE DESIGN CONSIDERATIONS

There are a number of specific problems with laboratory vacuum systems that must be addressed at the design stage that are not usually encountered in medical or industrial systems. These include what fluids are likely to be encountered such as water, VOCs, biological waste and particulate contamination to name a few as each system will have different variables, such as:

- Aggressive chemicals that may be drawn into the system.
- How large is the chamber holding the product that will be evacuated, e.g., a constant volume versus a continuous flow.
- How many users will be online at the same time.
- How long is an acceptable period for the evacuation process to take place.
- What will the final pressure in the evacuated chamber be.
- What safety measures must be taken to protect staff and users of the system.
- Could damage occur to the filtration system due to incorrect selection of materials of construction or incorrect filter elements.
- What method is used to determine how the removal of contaminants from the system will be accomplished, stored, and finally disposed of if necessary, the sewerage system may not be appropriate for many of these products and independent collection, storage, and removal may be required.
- Are there any additional design implications within the system that must be allowed for?
- What precautions have been included in the design to prevent the loss of vacuum due to use in remote locations within the system that may cause backflow in the pipeline? Backflow into previously evacuated equipment may cause cross contamination between samples.
- Should a VOC or aggressive chemical substance is drawn into the system that is incompatible with the vacuum pump or pipeline materials. What precautions have been built into the system design to prevent a system-wide failure?
- The incorrect operating pressures for the vacuum pump could damage the pump if it is operating outside the manufacturers recommended pressure settings.
- What measures have been taken to prevent an exhaust pipeline fire when working with a vacuum?
- If the vacuum pump is required to operate at pressures below −95 kPa calculations must be carried out to confirm the temperature of

the exhaust gas. The internal temperature of the pump and exhaust pipeline may require specific materials of construction with the addition of insulation to protect personnel from an overheated exhaust pipeline. If an oil sealed pump is used this will need to be taken into consideration as the superheated oil vapor could ignite in the exhaust chamber and further downstream in the exhaust pipeline. It may be necessary to consult with the manufacturer to confirm the pump type selected is suitable for the intended purpose and what if any additional safety features will be required.

The following formula is used as an example for the calculating the temperature of oxygen. The only variable that differs from the example is calculating the specific heat ratio for the exhaust gas which is an unknown. (if air is the likely medium the specific heat ratio is 1.4019).

The following is based on oxygen and is an extract provided with the permission of ASTM, taken from *ASTM G88–05* [14].

ASTM G88–05, 5.2.7.1 Compression Pressure Ratio [14] – In order to produce temperatures capable of igniting most materials in oxygen environments. A significant compression pressure ratio (P_f / P_i) is required, where the final pressure is significantly higher than the starting pressure.

NOTE 4—Equation 2.2 shows a formula (*the formula shown is based upon isentropic flow relations for an ideal gas. Source: ASTM, with permission.) for the theoretical maximum temperature (T_f) that can be developed when pressurizing a gas rapidly from one pressure and temperature to an elevated pressure without heat transfer.

$$T_f/T_i = [P_f/P_i]^{(n-1)/n} \tag{2}$$

where:

T_f = final temperature, abs;
T_i = initial temperature, abs;
P_f = final pressure, abs; and
P_i = initial pressure, abs.

$$n = C_p/C_v = 1.40 \text{ for oxygen} \tag{3}$$

where:

C_p = specific heat at constant pressure; and
C_v = specific heat at constant volume.

Additional implications for vacuum pump types, e.g., using a high vacuum pump for rough vacuum would place unnecessary demands on the pump, which may easily cause damage. While in operation to remove the bulk of the system gases from the pipeline before it begins to operate in its optimal design range. Using an oil seal vacuum pump to evacuate to a pressure of −99 kPa would be operating the pump at the extreme of its operating range.

2.3.7   VACUUM PUMP SELECTION CRITERIA FOR LABORATORY APPLICATIONS

The types of vacuum pumps for laboratory applications vary according to the pressure and temperature ranges as well as contaminants such as moisture and vapors from within the pipelines that are likely to be encountered during operation.

Many of the manufacturers of vacuum pumps provide similar styles and types of units so we should look at them as types rather than a manufacturer's models; each type has characteristics that meet a certain supply requirement. These vary widely and it can be broken into specific areas. The driving forces that will promote the purchase of one type over another include pressure, price, longevity, locality, and atmospheric conditions, local support, system design, flow requirements, and quality. We will look at each of these individually in this chapter. The following benchmarks are commonly used when selecting vacuum pumps for laboratory service.

2.3.7.1   PRESSURE

The pressure requirements for any vacuum plant is of primary importance, however, in the majority of laboratories, the maximum vacuum pipeline pressure is −60 kPaG (similar to medical vacuum pipelines) which is well within the operating pressure of most standard off the shelf vacuum pump types. It is common practice to provide package vacuum plants with built-in filtration. However, the available vacuum pressure from dry running rotary vane pumps may be insufficient for some laboratory applications.

2.3.7.2   VACUUM SYSTEMS (TO –99 KPA)

The design of laboratory vacuum plants, in general, has storage provided by the vacuum receiver, in addition to this, the pipeline will also provide some energy storage; However, this will be limited by the minimal differential pressure range available between the start and stop points of the pumps. A VSD controlled system has no differential operating pressure range and there is no requirement or need for a large vessel as storage is unnecessary in the system. A small vacuum vessel would be used to provide a buffer for any short-term pipeline pressure fluctuation. Providing vacuum pumps that are not used at their optimal pressure range may add to the cost of the equipment and the operating costs for no benefit.

The use of VSD controls will reduce the operating cost of any vacuum system as the efficiency of pumps working outside their peak efficiency pressure range decreases exponentially as the pipeline pressure approaches –101 kPa.

If the pump controls are using VSD's, the pressure selected for operation should be marginally below the desired pressure, i.e., if the desired pressure is –80 kPa the system operating pressure should be approximately –85 kPa. This may need final adjustment after commissioning to prevent the system from short cycling during use.

2.3.7.3   PRICE

The price of any vacuum pump is always a major concern. However, if the unit is being installed in harsh climatic conditions, it will be advisable to specify a superior quality plant, investing the additional money to provide additional protection, improved reliability, and serviceability for the future. Unfortunately, in many instances, budgetary constraints will override the above with future cost blowouts for repairs and maintenance plus the downtime for vital equipment and the associated losses that accompany it.

The selection of the correct vacuum pumps for the function should be based on the long-term operating costs plus maximizing features that provide real advantages f-or future repairs and maintenance.

2.3.7.4   LONGEVITY

The lifetime expectancy of vacuum pumps will depend on a number of factors:

- quality of the unit;
- location of the installation;
- maintenance provided and its availability; and
- the expected workload and running time.

The better the quality, the longer the unit will stand up to the rigors of hard use. The opposite to that is the supply of the best quality vacuum pump in a situation where it will operate minimally under good operating conditions with regular maintenance and service may be unnecessary and a reasonable quality vacuum pump will adequately provide this service. Unfortunately, the choice of equipment available in the marketplace may reduce the options available in this instance.

2.3.7.5   LOCALITY AND ATMOSPHERIC CONDITIONS

The site conditions will vary considerably from installation to installation. As an example, a plant being installed in a tropical environment in an open plant room will be subjected to extremes of humidity and temperature that an equivalent vacuum pump in cool climate location would not. Installation in a ventilated plant room may be necessary if the vacuum pumps, plant, and equipment selected have maximum or minimum temperature limitations. Water cooling plant is also an option that will need special consideration as water treatment will be necessary as well as the additional plant room floor space, cost, and safety concerns that go with it.

2.3.7.6   LOCAL SUPPORT

The ability of local service providers to maintain and service the plant will play a significant role in the selection of the manufacturer. Spare parts that need to be purchased from warehouses that are distant or overseas from the installation site are unacceptable; especially, if the installation is for research use. Operating staff relies on the ongoing supply being readily available, extended shutdowns while waiting for repairs or spare parts should be avoided at all costs. In some instances, it may be advisable to design multiple vacuum pumps and associated equipment or a supply of recommended spare parts to prevent this from occurring.

2.3.7.7   SYSTEM DESIGN AND FLOW REQUIREMENTS

The system designer must take all of the above into consideration when specifying any equipment. Flow rates are difficult to calculate as usually there will be no empirical data to call on and no documented references available. System pressures are suggested at –60 kPa; however, that may not be sufficient in some installations.

2.3.7.8   SITE CONDITIONS

When designing a laboratory vacuum plant, the selection of the vacuum pump type for the installation poses many variables. Considerations such as noise and vibration transmission to the building, heat rejection, power availability and floor space are just some that will need to be investigated during the design phase of the project.

2.3.8   VACUUM TREATMENT

2.3.8.1   MOISTURE REMOVAL

Moisture should not be drawn into the vacuum system under normal circumstances; however, this does occur regularly. Liquid traps should be provided at the connection points to the pipeline system or at the plant location prior to entry into the receiver or pumps. It may also be warranted to provide collection points at low points of the system. Should there be a concern that fluids could inadvertently be drawn into the system it may be advantageous to install local moisture separators with drainage available nearby.

Note: It may not be allowable to use the building drainage or sewerage system to dispose of this waste due to its nature however investigation should be undertaken to determine the correct method of disposal.

2.3.8.2   PARTICULATE AND BACTERIOLOGICAL CONTAMINATION REMOVAL

Entrained particulates can be removed by using biological filters that are available from the suppliers in the vacuum filtration equipment market-place. The standard method is to provide particulate filtration in the form of a bacteriological filter.

Bacteriological filters are installed prior to the vacuum receiver (where used) and the vacuum pumps. They remove particulates to 0.01 microns with an efficiency of 99.9995%. This filtration size will remove almost all known bacteria and some viruses. When removing and or replacing these filter elements, it should be standard procedure to wear protective clothing suitable for use with biohazardous materials and training for the end users should be a requirement of the project design documentation.

2.3.8.3   AGGRESSIVE VAPORS AND LIQUIDS SUCH AS VOCS

In certain installations, it will be necessary to protect the vacuum pumps from aggressive vapors and liquids if they are likely to be encountered in the system. Where highly volatile vapors with very low evaporation pressures (i.e., < −98 kPa) are encountered it is common practice to use local under bench vacuum pump sets with inbuilt condensers to extract these with back up mechanical filtration. These condensers can be connected to chilled water service if there is a need to reduce the temperature to ensure the efficiency of the condensation process. There are package sets for this currently available in the marketplace.

It may be necessary to use scrubbers to clean the exhaust from vacuum plants, these will need to be purpose-built or incorporated into compatible mechanical plant systems. The vacuum pump manufacturer would be the ideal place to begin an investigation, if this is required as they will need to provide guarantees that their equipment would be suitable for the service and also advise on what they require to maintain it and for warranty and maintenance guarantees.

Note: Biological, flammable and toxic exhaust should not be vented into scrubber systems unless specifically designed for this purpose.

2.3.8.4 BACTERIOLOGICAL FILTRATION

The majority of instrument air filter manufacturers provide vacuum filters designed for use in medical vacuum pipeline service. These filters use coalescing elements that will eliminate particulates down to 0.01 microns.

The filters' aluminum housings are manufactured from aluminum with a glass fiber element (e.g., oleo phobic borosilicate or equivalent). It may be necessary to provide a higher level of protection from some aggressive solvents; if necessary, there are housings available using the same glass fiber elements with stainless steel casing and filter casings to ensure compatibility with the contaminants. However, the manufacturer must be contacted to confirm that the filter as a whole is capable of meeting the system demands.

Bacterial filters are used on the vacuum pipeline to prevent any microbial or particulate contamination from entering the vacuum pump or vessel and protect the exhaust from contamination. It is not possible to protect the plant from moisture contamination using these filters and the use of drainage or collection traps. At the plant location plus provision of collection points at the lowest point of any pipeline risers or at the base of the main supply pipeline in a multi-floor supply riser, may be an advisable inclusion.

It may be advisable to include activated carbon vacuum filters where odors or other contaminants may be encountered; especially, if hydrocarbon vapors, VOCs or similar contaminants need to be extracted from the air stream. Consultation with the manufacturers is highly recommended.

2.3.8.5 VACUUM EXHAUST

The exhaust gases leaving the vacuum pumps may be at elevated temperatures due to the compression occurring during pump operation. At extreme vacuum pressures the temperature is sufficiently elevated to create a physical hazard should it come into contact with staff or service personnel. It may be necessary to provide insulation of the exhaust pipeline for the full length to the termination, preferably at the building exterior. The exhaust temperature can be calculated using Gay Lussac's Law. The evacuated air may also contain contaminants including moisture, evaporated VOCs, particulates (e.g., dust), gases (e.g., CO) and microbes that have not been captured by the pipeline filtration system and it may be necessary to remove these contaminants from the exhaust gases prior to venting to the atmosphere.

If there is any concern that the vacuum pump exhaust may carry hazardous contaminants, it is highly recommended to provide filtration in the exhaust pipeline from the vacuum pumps based on the expected level and type of contamination. Discussions with the laboratory manager and the filter supplier to determine the suitability of the proposed filters, to ensure exhaust compatibility, are necessary.

Exhaust filters must be sized to suit the system design flow rate at free air conditions with the minimum back pressure allowable to meet the vacuum pump specification. Excess back pressure will cause loss of pump performance. This may be due to restriction in the filter and must be below the maximum tolerances and in accordance with the pump manufacturer's recommendations.

2.4   CRYOGENIC STORAGE

The storage of large volumes of gas presents a number of issues that must be taken into consideration when designing the storage facility for the supply of gases to the laboratory. In most laboratories, the three gases usually requiring a supply of large volumes of gas are nitrogen and argon and to a lesser extent oxygen and helium.

The major problems facing designers include allocating sufficient space to store cylinders both full and empty or the location of a cryogenic storage site for a bulk supply. Both of these will usually require compliance with national government and or local regulations. Incorporated in the site will be the provision of lighting, power supplies for operating the cryogenic tanker's cryogenic pump as well as telemetry or some form of monitoring the content of the vessel or vessels during operation.

Delivery access for cylinder delivery or filling the cryogenic vessel will also be a necessity including a secure concrete hardstand for a tanker if oxidizing gases are being supplied. The size of the storage site should be discussed with the client or with proposed suppliers, if they are available. It is also advisable to allow sufficient space for expansion for future use if demands are likely to increase which would include a concrete slab able to support the weight of a larger vessel (Table 2.1).

TABLE 2.1 Comparative Volumes of Liquid to the Gaseous States and Numbers of Cylinders

Liquid Vessel Volume in liters	Gaseous Argon in liters	Gaseous Nitrogen in liters	Gaseous Oxygen in liters
3,000	2,472,000 liters	2,046,000 liters	2,529,000 liters
	or	or	or
	353 G size cylinders	292 G size cylinders	361 G size cylinders
5,000	4,120,000 liters	3,410,000 liters	4,215,000 liters
	or	or	or
	589 G size cylinders	487 G size cylinders	602 G size cylinders
10,000	8,240,000 liters	6,820,000 liters	8,430,000 liters
	or	or	or
	1,177 G size cylinders	974 G size cylinders	1,204 G size cylinders
15,000	12,360,000 liters	10,230,000 liters	12,645,000 liters
	or	or	or
	1,766 G size cylinders	1,461 G size cylinders	1,806 G size cylinders
20,000	16,480,000 liters	13,640,000 liters	16,860,000 liters
	or	or	or
	2,354 G size cylinders	1,949 G size cylinders	2,409 G size cylinders
25,000	20,600,000 liters	17,050,000 liters	21,075,000 liters
	or	or	or
	2,943 G size cylinders	2,436 G size cylinders	3,011 G size cylinders
30,000	24,720,000 liters	20,460,000 liters	25,290,000 liters
	or	or	or
	3,531 G size cylinders	2,923 G size cylinders	3,613 G size cylinders

Note: a "G" size cylinder at a pressure of 14,000 kPa (2,000 psi) contains the equivalent of 7,000 liters or 250 cubic feet of gas, cylinders with higher pressures will contain considerably more gas. Carbon dioxide and nitrous oxide are the exceptions to this rule as they are stored as liquids at ambient temperature.

The use of cryogenic gas supplies circumvent some of the storage issues for these gases. The space required by liquid vessels to store large gas volumes reduces storage space dramatically when compared to cylinders as can be seen from Table 2.1.

As an example, it is possible to store up to 30,000 liters of liquid oxygen on a relatively small concrete slab whereas to store the same volume of gas in cylinders would represent 3,613 x 7,000-liters cylinders in its gaseous state. A liquid gas storage vessel is a vacuum jacketed vessel; the inner vessel manufactured from stainless steel and the outer from mild steel. The interstitial space is filled with Perlite or specially developed in-house insulation products and then evacuated to pressures around 10^{-8} kPaA.

To convert this liquid to its gaseous state it is necessary to pass the cryogenic liquid through a heat exchanger, the heat exchanger would normally be an ambient vaporizer designed to convert the cryogenic liquid to its gaseous state to meet the required volumes as required by the laboratory to satisfy system demand. The size of the heat exchanger will be dependent on the expected flows from the vessel. It is necessary to discuss this with the gas supplier as individual gas manufacturers may use different types of heat exchangers that will vary in size and format.

As the liquid level drops in the vessel, the head pressure needs to be maintained at the predetermined vessel pressure. Any variation in this pressure may affect the system pipeline and the ability of the pipeline to meet the flow and pressure demand requirements of the laboratory. The internal pressure is maintained by the pressure building coil usually mounted beneath the vessel itself. It draws a small amount of liquid from the bottom of the vessel that is vaporized and then returned to the vessel through internal pipe connections. It is self-regulating and controls the internal pressure at a predetermined level.

In many countries, the gas suppliers lease the cryogenic vessels due to the manufacturing cost plus what would otherwise be extended manufacturing times and deliveries. The associated vaporizers and pressure control equipment may also be leased to the laboratories as part of the gas supply contract. The vessel leasing arrangements are usually under direct negotiation between the gas supplier and the laboratory or through government contract. Site installation of the vessel and the associated controls are usually carried out by the gas supply company during the final stages of construction. The supplier will calculate the size of the vessel in accordance with the system demand taking into account their delivery schedules and the relative location of the site and the distance to their storage and supply depots.

Site storage for these vessels is usually regulated by National or Regional Authorities who provide details on acceptable locations with respect to access of the public and safety measures that are deemed appropriate when working with cryogenic gases.

Information about the size and construction details for concrete hard stands and any associated safety requirements are available from the gas supply companies.

It may be necessary to obtain certification from the national authority or similar organization in most countries prior to installing a cryogenic vessel on any site.

Liquid gases have special conditions that require individual consideration. In their liquid state, these gases are stored at cryogenic temperatures (below $-150°C$), oxygen at $-183°C$, Argon at $-186°C$, Nitrogen at $-196°C$, and helium at $-269°C$. The liquid is converted to gas at predetermined pressures for supply to the laboratory, this is accomplished in two steps. Firstly in the vessel that must maintain a constant internal head pressure, usually at around 1,000 kPa although this is a variable and can be pre-set to suit the laboratories requirements. And secondly, conversion of the liquid for supply to the facility. Each of these two processes is carried out independently. Individually controlled vaporizers or heat exchangers for each are required to convert the liquid to its gaseous state. Subsequent to the liquid being converted to gas the pressure may need to be reduced in pressure to meet the special requirements the laboratory may have or in some instances the gas may remain at vessel pressure for reticulation throughout the facility with local pressure reducing stations.

The vessel's internal pressure is controlled by converting small amounts of liquid into gas using an independent pressure building coil. This is usually located beneath the vessel and provides a 1,000 kPa pressure controlled supply to the vessel.

The supply to the laboratory is drawn from the bottom of the vessel as liquid and converted to gas through independent ambient heat exchangers that supply the gas on demand at the pre-set pressure prior to the main pipeline supply to the facility.

The use of cryogenic liquids is common in many laboratories and specialist knowledge is necessary prior to the commencement of any cryogenic plant and pipeline design.

If the laboratory requires a cryogenic liquid, it is taken from the bottom of the vessel and piped to the point of use. The pipeline can be manufactured from copper with a polyurethane insulation or using stainless steel with a stainless steel internal pipe with a pressure-sealed vacuum jacketed outer casing.

The difference between the two systems is the efficiency of the insulation. No matter which system is selected, the pipeline will continuously adsorb heat from the atmosphere causing the cryogenic liquid to convert to its gaseous phase. Where pipelines are installed for continuous service, the use of vacuum jacketed pipelines are recommended to provide the most efficient insulation and thereby prevent the continuous loss of liquid through evaporation caused by heat transference to the internal pipeline. An example of this type of use would be an MBE that demands continuous cryogenic nitrogen supply for extended periods with a minimal gaseous content.

Polyurethane insulation may be suitable for systems that are not continuously in use and the cryogenic liquid is not in demand at short notice; examples would be for filling of portable dewars that may only be used irregularly.

Note: The cost difference between a copper pipeline with polyurethane insulation and a stainless steel vacuum jacketed insulation is considerable.

If the system requires a constant supply of gas-free liquid, it is necessary to provide a means of removing the evaporated cryogenic liquid in its gaseous state from the site of the local application, as well as from the pipeline. This is achieved using a phase separator connected at a high point in the pipeline and as close to the point of use as possible. The phase separator effectively separates the liquid and gas and vents the waste gas from this high point in the pipeline. This gas is piped to a safe location away from the building preferably at a high level. It must be remembered that this exhaust gas is at a similar temperature to the liquid and the exhaust pipeline should be manufactured from the same material as the pipeline. Failure to provide suitable insulation will cause the formation of large volumes of ice on the outer casing of the pipeline that over time may increase in size and may cause water and ice damage to building framework and furniture that may be nearby.

An additional concern is at the termination of this exhaust pipeline where the gas is vented to atmosphere, the contact point with the atmosphere. The waste gas at close to cryogenic temperatures will cause a build-up of ice around the exhaust port. To lessen this effect cryogenic pipeline manufacturers may be able to provide electronic heaters that reduce the build-up and are to be recommended for use when exhausting these gases.

Stainless steel vacuum jacketed pipelines are generally manufactured in sections of various lengths which are evacuated and sealed prior to delivery to the laboratory. The vacuum level in these sections is in the vicinity of 10^{-8} kPaA and great care should be taken when assembling the sections and any valves and associated equipment.

The pipeline design will necessarily include changes of direction, isolation valves, solenoid valves, relief valves, and other system controls. These are available from the manufacturers of the stainless steel vacuum jacketed pipelines and are supplied with the same insulation as the pipeline. It is advisable to contact the manufacturers of this equipment to ensure that the valves and equipment required for the pipeline are available at the working pressures that are necessary, as well as being suitable for the proposed purposes.

The various pipe sections, changes of direction and any vacuum jacketed valves, are all connected using a bayonet type coupling that will require space for assembly. These fittings are not flexible and require a suitable distance to allow insertion of the bayonet connection into the sections of the pipeline.

Points that need to be taken into consideration when designing a cryogenic pipeline include:

- The design of a cryogenic system must be carried out by a suitably qualified and experienced engineer with a complete understanding of the characteristics of the liquids being piped. Without this knowledge, it is not possible to provide a safe working system or to begin designing any cryogenic pipeline.
- An extensive knowledge of the proposed materials of construction, valves, fittings, and controls likely to be required is an essential prerequisite. This information is only available from the manufacturers of the equipment being used.
- An understanding of the occupational health and safety aspects plus local regulations, laws, and site-specific requirements that will be encountered by the installers, operators, and personnel who will be using the system.
- The pipeline layout must make allowances for the installation of rigid sections of pipe, bends, valves, and other control equipment.
- It may be necessary to allow a rise or fall to points of the pipeline to assist in the flow of the liquid and for connection of the phase separators if required.
- The phase separator must be designed into the system taking into consideration where it is proposed to be located with sufficient

space above the pipeline to accommodate the unit. Phase separators require headspace above the pipeline and as these are proprietary items they cannot be manufactured to fit specific locations.

- Phase separators may also have independent pressure controls built into the unit. The pressure requirements of the end user must be ascertained prior to specifying the equipment and configuration.
- Exhaust cryogenic gases must be removed from the work area either by direct extraction or through the collection and venting through the phase separator. This should be discussed with the proposed manufacturer of the equipment using the liquid and the manufacturer of the pipeline.
- It may be necessary to provide a pressurized gas supply or instrument air or gaseous nitrogen supply for control purposes. The manufacturer will be able to provide this information and specific pressure and flow requirements.
- A relief valve must be installed in every section of pipe where possible isolation between valves or terminations of pipelines may occur. Trapped cryogenic liquids will eventually be warmed to room temperature and will expand to their original volume which within a confined space may increase in pressure to in excess of six hundred bar. This would rupture any cryogenic pipelines.
- Any relief valves fitted inside the laboratory and connected to the cryogenic system must be exhausted to the building exterior. A valve failure during operation must take into consideration the volume of gas that could be vented into the laboratory.
- Gas sensing and monitoring systems should be installed in any location where cryogenic systems are being installed.

Non-vacuum jacketed valves and controls are made of brass and manufactured by major valve suppliers and are readily available. These are manufactured with extended spindles on the handle stem that can be encased in polyurethane in a similar manner to the pipeline. Standard practice is to encase the pipeline in PVC tubing with internal supports in the interstitial space that prevent the internal pipeline from contacting the PVC tube wall. When the pipeline is complete, holes are drilled in the PVC tube and a polyurethane liquid mix is poured into the interstitial space. As the liquid warms it will expand and fill the space and provide a level of insulation suitable for this type of installation. This is not a recommended system for cryogenic gases that are connected to phase separators and are expected to be full of liquid continuously as the gas loss caused by heat in the leak, due to the inefficiency of the polyurethane insulation compared to vacuum insulation, will be considerable.

Note: Polystyrene is not suitable for use with cryogenic liquids due to its minimum working temperature of approximately –60°C.

2.4.1 LIQUID NITROGEN

The supply of cryogenic liquids is provided through two major pipeline styles:

a) Stainless steel vacuum jacketed insulation available in static or dynamic types and copper pipelines with polyurethane insulation.

b) Liquid nitrogen is used for cooling of laboratory equipment such as MBEs and to fill cryostats or portable dewars used for the storage of cells or similar at cryogenic temperatures.

The liquid in both cases is stored at a temperature of –196°C, the VIE or VIV stores the liquid at a pressure of approximately 1,000 kPa (this varies between facilities and suppliers).

The fixed, mobile or portable laboratory container, called a cryostat or dewar, is a non-pressurized vacuum insulated vessel that allows the liquid to evaporate and vents the gas vapor into the local environment. Evaporation rates for cryogenic liquids for VIE/VIV's are approximately 0.4% per day for large vessels through to 1.0% per day for smaller vessels. Dewars are variable in size and function and the manufacturer should be contacted to provide information as they vary in use from supplier to supplier although they are rarely supplied as part of an installation. In any location where liquid nitrogen is stored in a closed location (storeroom or laboratory even if ventilated). It should include a gas monitoring system with audible and visual alarms located inside and externally to the area including a manual activation point within the storage area preferably near the exit.

Cryogenic vessels may be used for both liquid and gaseous supply, both of which may be piped directly to the point of use. These could include fixed automatic cryostats that provide the liquid nitrogen supply on demand or to specialized laboratory equipment such as a molecular beam epitaxy (MBE). Cryogenic vessels have the option for a gaseous supply where the liquid is passed through a vaporizer or heat exchanger for distribution to the laboratory.

2.4.2 LIQUID ARGON

The supply of liquid argon is provided only for gaseous use where large volumes of argon are required.

The liquid is stored at a temperature of –190°C, the VIE or VIV stores the liquid at a pressure of approximately 1,000 kPa (this varies between locations and suppliers).

Cryogenic vessels are used for gaseous argon withdrawal only, which is piped directly to the point of use in the same manner as a cylinder supply.

2.4.3 LIQUID OXYGEN

The supply of liquid oxygen is provided only for gaseous use where large volumes of oxygen are required.

The liquid is stored at a temperature of –183°C, the VIE or VIV stores the liquid at a pressure of approximately 1,000 kPa (this varies between locations and suppliers).

Cryogenic vessels are used for gaseous oxygen withdrawal only, which is piped directly to the point of use in the same manner as a cylinder supply.

2.4.4 LIQUID HELIUM

The supply of liquid helium is provided only for gaseous use where large volumes of helium are required.

The liquid is stored at a temperature of –269°C, the VIE or VIV stores the liquid at a pressure of approximately 1,000 kPa (this varies between locations and suppliers).

Stainless steel vacuum jacketed insulation is recommended for use with liquid helium and is available in static or dynamic types.

Cryogenic vessels may be used for both liquid and gaseous withdrawal, both of which may be piped directly to the point of use to supply the liquid helium supply on demand or to specialized laboratory equipment. Cryogenic vessels have the option for a gaseous supply where the liquid is passed through a vaporizer or heat exchanger for distribution to the laboratory.

Many laboratories incorporate helium recovery systems that are manufactured by specialist companies. The processes include the recovery of the waste helium, storage, purification, recompression, and liquefaction of the gas. It is not the intention of this text to cover this subject as it is usually provided by specialist suppliers.

2.4.5 LIQUID CARBON DIOXIDE

Liquid carbon dioxide is not recognized as a cryogenic liquid due to its operating temperatures and is available in two formats; high-pressure liquid at a pressure of 5,824 kPa at ambient temperature supplied from a high-pressure gas cylinder. It is also available as a refrigerated liquid supplied in a cryogenic vessel or refrigerated in an insulated storage vessel.

High-pressure liquid carbon dioxide is used directly from the gas cylinder at a pressure of 5,824 kPa for specific gas chromatographs where it provides cooling to the chamber by evaporating a quantity of liquid under high pressure. The method of supply is a high-pressure cylinder that draws liquid from the bottom of the cylinder direct to the GC through a tube connected to the cylinder valve that allows withdrawal of the high-pressure liquid.

It is not possible to regulate the pressure of this liquid for this purpose, it is used at full cylinder pressure of 5,824 kPa. When using this system in a pipeline it is recommended that an "ullage" vessel is installed above the highest point of the pipeline to provide a gas head. This allows provision for expansion due to local ambient temperature conditions increasing during periods of non-use should the high-pressure liquid be unable to return to the high-pressure cylinder.

Note: Liquid carbon dioxide from a cryogenic or refrigerated storage vessel is not suitable for this purpose as the pressure is considerably lower than that necessary for use in a GC.

2.4.6 CRYOGENIC AND REFRIGERATED STORAGE VESSELS

Cryogenic and refrigerated storage vessels are usually provided on lease from the gas supply companies, they are available in a range of sizes to suit the design flow rate of the system and the availability of regular supply which may be impacted by the location of the facility and the supplier's delivery that may not be readily available in remote locations. It is recommended to allow the gas supplier to arrange the vessel details with the facility; however, it will be necessary to provide site facilities to support the vessel, and provide adequate off-street parking for the delivery vehicle as well as the necessary services allow monitoring of the vessel contents.

The site storage of cryogenic liquids may be regulated by national authorities and compliance and approval for the location must be obtained prior to commencing construction.

The vessel site will require the following services plus contact should be made with the gas supplier to coordinate any additional requirements they have:

- The concrete slab where the vessel will be located must be constructed to support the vessel when full of liquid, the vessels themselves plus the liquid have a substantial weight and should be designed by a qualified structural engineer. The gas supplier should be asked for information about the size of the concrete slab and also may be able to provide further information about the construction requirements to suit the specific vessel size.
- Additional area for the evaporator plus supply of emergency backup supply cylinders, this should be discussed with the architect and the gas supplier.

2.4.7 MECHANICAL SERVICES REQUIRED

- Power supply to operate the cryogenic pump on the delivery tanker.
- Power supply and local lighting for use when access is required after hours.
- Water supply for washing down the area as necessary.
- Security fencing.
- Telemetry connections if the gas supplier has the option for remote monitoring of the vessel and its contents.
- Wiring connection from the facility gas monitoring alarm system.
- Laboratory gas, and if required an insulated liquid pipeline connection; the gas supplier may require that final connection to the cryogenic vessel are carried out by their own staff.

- A remote filling connection may be if the delivery vehicle cannot park adjacent to the cryogenic vessel.

A pressure control station that reduces the supply pressure from the vessel to the pipeline pressure will be necessary. This may be supplied as part of the gas supplier's scope of work or may be part of the pipeline construction. The panel may be installed at the cryogenic vessel location or in a remote situation closer to the point of use as may be dictated by the project prerequisites (Figure 2.15).

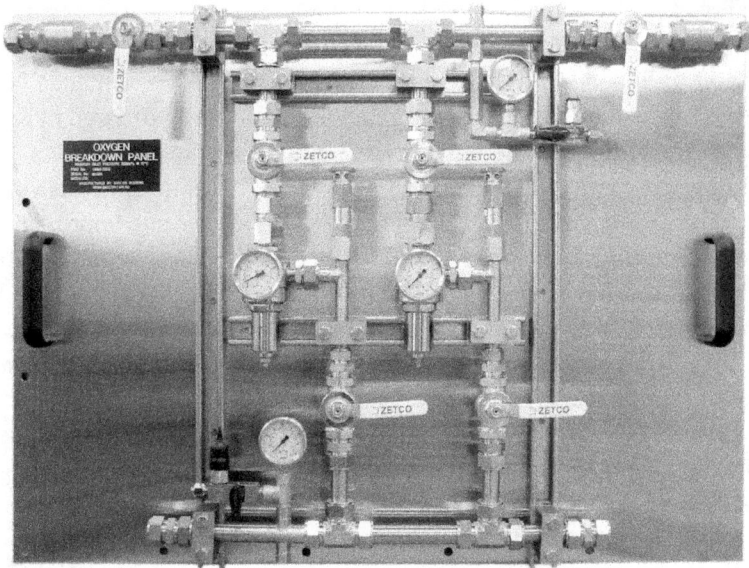

FIGURE 2.15 Gascon cryogenic vessel pressure control panel.
(**Source:** Gascon Australia. With permission.)

2.5 GAS MANIFOLD SYSTEMS

Cylinder manifolds provide continuous gas supply in automatic, manual, and single cylinder configurations, these manifold connections are suitable for large multi-cylinder manifolds with independent banks of cylinders.

The selection of the correct cylinder supply arrangement should be based on the projected usage, the type of gas required and the location of the supply source. It may also be necessary to comply with local national regulations governing the storage of flammable, toxic, inert, and high-pressure gas cylinders.

The supply of all gases are available as cylinders with a variety of purity levels which is suitable for small demand systems or very specialized gas supplies; however, for large institutions, it is necessary to provide bulk supplies that gas cylinders cannot meet.

The supply of cryogenic gas supplies for smaller installations may be available in a portable vacuum insulated vessels and is limited to argon, nitrogen, oxygen, and refrigerated carbon dioxide which may be suitable for facilities that require constant long-term demand systems using regulated flows.

This chapter is specifically looking at the supply of gases using manifold cylinders with sufficient numbers to meet the demands of the system. The volume of laboratory gases in the majority of large cylinders is approximately 7,000 liters; however, higher pressure supplies are available for some gases with increased volume. The exceptions to this volume are carbon dioxide and nitrous oxide both of which contain approximately 14,000 liters of gas which are stored in their liquid phase under high pressures (refer to Chapter 5).

Note: There is a variation in the volume of gas contained in cylinders of similar sizes due to different dimensions of the cylinders from diverse manufacturers and the use of a variety of materials used in cylinder manufacture.

Manifolds are differentiated into general types, fully automatic, semi-automatic, manual, and single cylinder supplies each of which is designed to meet specific design parameters.

2.5.1   AUTOMATIC MANIFOLDS

The Gascon auto change-over manifold consists of a pressure control assembly made up of three separate pressure regulators – two first stage regulators and a second stage pipeline regulator. On the first stage regulators, the RHS cylinder supply bank is connected to the rear regulator and the LHS cylinder supply bank is connected to the front regulator. The first stage regulators reduce the cylinder pressure to a lower intermediate pressure. A cylinder bank selection lever simultaneously adjusts pressure setting on both the front and rear regulators (increasing the pressure on one while decreasing the pressure on the other by operating the control handle). The direction in which the lever is pointing controls which cylinder supply bank the manifold will start drawing gas from. The cylinder supply bank to which the lever is pointing towards is called the "IN USE" bank, and the other bank is called the "RESERVE" bank (Figure 2.16).

FIGURE 2.16 Gascon automatic changeover manifold.
(**Source:** Gascon Australia. With permission.)

When the "IN USE" bank empties, the pressure differential between the two first stage regulators causes the manifold to automatically start drawing gas from the "RESERVE" bank. (Note, the supply bank selection lever does not move). If required, pressure switches can be used to generate a signal to indicate that a change-over has occurred. The empty cylinders should be replaced as soon as possible. Inlet non-return valves (NRV) on either side of the first stage regulators prevent gas from decanting from one cylinder bank to the other cylinder bank.

To replace the recently emptied cylinders, first, move the lever towards the "RESERVE" bank that the gas is now being drawn from. After moving the lever, the "RESERVE" bank now becomes the new "IN USE" bank. Replace the emptied cylinder as detailed in the operating instruction. It is

important to alternate the "IN USE" and "RESERVE" banks in this procedure. Simply replacing the emptied cylinder(s) and not moving the lever
will mean that the gas supply in "RESERVE" bank will slowly empty to a
stage that eventually there will be no "RESERVE" gas supply.

After the first stage regulators, the gas passes through the second stage/
line regulator that reduces the pressure to the final pipeline pressure.

A modular inlet header extension is used to allow multiple cylinders
or packs of cylinders to be fitted to either side of the automatic manifold.
They also allow for the storage capacity of the automatic manifold to
increase in the future.

Manifolds should incorporate the following capabilities:

- Semi-automatic or fully automatic cylinder changeover system
  suitable for the purity of the gas being connected.
- Pressure gauges to indicate the current cylinder contents of both
  banks of cylinders.
  Note: Carbon dioxide and nitrous oxide cylinders do not change
  pressure until the gas which is stored in its liquid high-pressure
  state is exhausted. At that time the remaining cylinder contents is in
  its gaseous state and will be used rapidly.
- Pressure gauge to indicate the regulated pipeline pressure.
- Indication of the "IN USE" and "RESERVE" operating cylinder
  banks.
- Pressure switches or transducers for detecting cylinder change-over
  and line pressure failure.
- Test port isolation.
- Purge ports for both manifold cylinder banks to remove impure gas
  from the flexible cylinder connections prior to opening the newly
  replaced cylinder bank to the pipeline.
- High-pressure flexible leads for cylinder connection with suitable
  CGA connections.
- Cylinder retaining brackets.
- Facility to provide storage for spare gas cylinders usually equal to
  the number of cylinders connected to one bank usually attached to
  the manifold.

Cylinder supply manifolds are manufactured as gas specific items and
cannot be altered to connect to or to supply alternative gases. They are
available in a range of capabilities based on the gas type. The pipeline

pressure and flow requirements of the system which must be determined at the design phase of the project.

Note: Semi-automatic manifolds require manual resetting after the automatic changeover mechanism has operated and the gas monitoring alarm system has been activated. The manual reset reverses the pressure settings that activate the "IN USE" and "RESERVE" functions of the manifold. The reset procedure is carried out after the replacement cylinders have been connected to the manifold which ensures that the "RESERVE" cylinders are always the most recently connected full cylinders and the "IN USE" cylinders are the cylinders that have been in operation prior to the replacement, are partially depleted, and are used first.

2.5.2 MANUAL MANIFOLDS

Manual manifolds consist of two banks of cylinders or cylinder packs consisting of one or more cylinders on each side. The independent cylinder banks are controlled manually by the use of isolation valves connected to the manifold which must be manually operated to select the "IN USE" and "RESERVE" cylinder banks (Figure 2.17).

FIGURE 2.17 Gascon manual changeover manifold with purge valves fitted. (**Source:** Gascon Australia. With permission.)

This type of manifold is used where physical monitoring of the contents of the banks of cylinders is easily provided to monitor gas use.

During operation, the manifold uses gas from the "IN USE" cylinder bank until it is depleted at which time it must be manually changed to the "RESERVE" bank.

The selection of ancillary equipment such as the pressure regulator, pressure relief valves, flashback arrestors, solenoid shutoff valves, and pressure switches for alarm monitoring should all be fitted either as part of or adjacent to the manifold.

2.5.3   SINGLE CYLINDER MANIFOLDS

Single cylinder manifolds consist of a single cylinder. The cylinder is manually monitored and is the only "IN USE" cylinder (Figure 2.18).

FIGURE 2.18   Gascon single cylinder connection.

(**Source:** Gascon Australia. With permission.)

This type of manifold is used where low usage is likely and continuous monitoring of the contents of the cylinders is available.

The selection of ancillary equipment such as the pressure regulator, pressure relief valves, flashback arrestors, solenoid shutoff valves, and pressure switches for alarm monitoring should all be fitted either as part of or adjacent to the manifold and must be detailed in the specification including materials of construction, purity level, pipeline pressure and type and settings for any equipment fitted.

2.5.4 LABORATORY GAS ALARM SYSTEM

Laboratory gas alarms are usually provided in locations where they can be visually observed easily and regularly. They should be independent of other alarm panels in the system and the removal of any panel must not interfere with the operation of the system.

All alarm panels may be interconnected with selected alarm messages or alarms repeated at complementary locations to ensure any major alarm function is able to be acted upon.

Alarm panels should incorporate functions that indicate the integrity of the wiring system, either short or open circuit, between the switching devices and the alarm panels themselves. The system should be interfaced with any building BMS if available.

Local alarm panels should incorporate all or some alarms for each gas directly connected to the laboratory gas systems and be able to indicate selected messages from sources such as the gas supply manifolds, instrument air, and suction plant. There should be a central alarm panel programmed to indicate all messages from all panels in the laboratory, the location of which should be in a central location that is continuously monitored.

It is common practice to use the same equipment available for medical gas monitoring. These alarms include all of the necessary functions for monitoring the gas supply manifolds and plant and may also accept alarm functions from other building services. Where possible the use of alarm systems with an IP addresses interface ability should be selected. This will allow interconnection to the BMS, local mobile telephone services and the LAN if available.

2.5.5   FLAMMABLE AND HAZARDOUS GAS ENVIRONMENTS

When pressure switches are required for manifolds using flammable gases or gas mixtures containing flammable components. The manifold must be segregated and located in an isolated flammable or toxic gas environment determined by national or similar National or Regional Authorities, hazardous and toxic atmosphere guidelines must be followed. The switches must be either intrinsically safe or the signal to and from the switches needs to be conditioned to be intrinsically safe by using Zener barriers, Transformer Isolation Barriers or other similar devices. Different flammable and toxic gases may require different levels of protection. Consult local authorities and the manufacturer of the manifolds and switches for further details.

2.5.6   FLASHBACK ARRESTORS

A flashback arrestor is required for acetylene, hydrogen, and carbon monoxide or other flammable gas systems. A flashback arrestor is recommended especially where the gas has the possibility of mixing with an oxidizing gas. Many flammable gas flashback arrestors have a relatively low maximum working pressure; typically around 100 to 350 kPa depending on the gas type which is often well below the required pipeline pressure. It is worthwhile to double-check the pressure rating of the flashback arrestor to ensure its suitability before installing them into a system.

Flashback arrestors for oxidizing gases are recommended when that gas is used in conjunction with an oxidizing gas and has a possibility of mixing with a flammable gas such as in an AAS. Flashback arrestors are not required for inert gases.

2.6   LABORATORY CYLINDER STORAGE CABINETS

The provision of local cylinder supplies within a laboratory may be a necessary requirement in some instances due to the specialized nature of some rare gases, e.g., neon and krypton, which poses specific storage difficulties and safety concerns. It is possible to provide local storage cabinets that include protection of the cylinders and manifold controls, internal cabinet monitoring; exhausting of vented and escaped gases and continuous flushing of the cabinet interior.

All the following information in Section 2.6 is provided by DÜPER-THAL Sicherheitstechnik GmbH & Co. KG: Duperthal's Supreme line of safety storage cabinets are ideal for the storage, use, and emptying of pressurized gas cylinders. The supreme line features a fire resistance of 90 minutes for a temperature increase of 50 Kelvin measured at the neck of the gas cylinder. Optimized interior height enables maximum access during installation and operation of the gas cylinder fittings. The cabinet is equipped with a rolling ramp for easy access for the installation and removal of the cylinders. The standard integrated installation rails are adjustable in height for convenience. Pressurized gas cylinders can be secured against accidentally falling over by means of standard integrated cylinder holder with retaining belts. Pipes and cables can be directly laid from the gas fitting through the cabinet ceiling to the outside.

With a fire resistance of 90 minutes, the Type G90 range can prevent the spread of a fire within the laboratory and potentially save lives within 90 minutes; staff can be evacuated and fire-fighters can tend to the hazard. Constructed from high-quality powder-coated sheet steel, the Duperthal Type G90 storage cabinets have smooth cabinet surfaces with no protruding hinges or covers.

The cabinet inclusions are:

- Fire resistance of 90 minutes for a temperature increase of 50°C.
- Third-party Certified and approved by TÜV-certification body.
- Wing doors sheet steel, high resistance powder coated.
- Outer carcass sheet steel, high resistance powder coated.
- Inner carcass made from high-quality décor panels.
- Two installation rails for holding gas fittings.
- One cylinder holder each with one retaining belt per standing space.
- Integrated drilling template.
- Standard integrated installation rails adjustable in height.
- Rolling ramp.
- Ventilation and extraction in the whole cabinet due to slits in the air supply and exhaust air ducts.
- Door opening angle continuously adjustable up to 170 degrees.
- Exceeds requirements of Australian Standard AS/NZS 1940.
- Type-tested and classified to European standards DIN EN 14470–2 and DIN EN 14727.

The DUPERTHAL safety cabinets are approved for the storage of pressurized gas cylinders in working spaces in accordance with the industrial safety technical rules on the storage of hazardous substances in portable containers (TRGS 510:2013 Annex 3). The internal height of the safety cabinet ensures that standard commercial pressurized gas cylinders including the necessary pressurized gas fittings can be placed in the cabinet and fitted.

The volume of all pressurized gas cylinders in safety cabinets may not exceed 220 liters. This includes pressurized gas cylinders for purging gases with a maximum volume of 10 liters. Depending on the model, up to 4 pressurized gas cylinders size 50 liter or up to 3 pressurized gas cylinders size 70 liter can be stored in a safety cabinet.

Prohibited use: The safety devices and features installed in the DUPERTHAL safety cabinet, e.g., shut-off dampers in the air ducts, interlocking elements, etc., must be in full working order and kept so at all times.

Prohibited use is also deemed to be granting unqualified or inadequately trained personnel and unauthorized persons free access to the hazardous substances stored in the safety cabinet.

It is also prohibited for the owner/operator of the safety cabinet to fail to fulfill their statutory obligations, e.g.

- Failure to issue the required in-house instructions for use; and/or
- Failure to perform the risk assessment and to prepare the explosion protection document;
- Failure to carry out the specified maintenance and inspection work and recurring tests; and/or
- Failure to instruct the personnel in proper storage and removal at least once a year, with written documentation of the instruction given.

The risks in the event of failure to observe the safety instructions are:

- Failure to observe the safety instructions can result both in risk to people as well as the environment and technical equipment.
- Failure to observe the safety instructions can result in all guarantees and compensation claims becoming null and void.
- The safety instructions are given in these operating and maintenance instructions; the existing legal regulations as well as the accident.

- Prevention regulations and the owner/operator's company and safety regulations shall be observed.

The safety instructions to be provided for the owner/operator are:

- The restrictions on storing hazardous substances and preparations together are to be observed.
- The owner/operator shall forbid unauthorized persons to access the hazardous substances stored in the safety cabinet.
- The owner/operator shall regulate smoking, handling, and working with naked flames in and on the safety cabinet.
- Safety devices are to be kept fully functional at all times.
- The function of the inlet and air extraction openings of the industrial venting shall not be impaired.
- Safety instructions for maintenance, inspection, and installation work.
- The owner/operator shall ensure that all maintenance and installation work is carried out by authorized, qualified personnel.
- All work on the electrical installations shall only be carried out by qualified electricians when the system is off-loaded. See also the relevant accident prevention regulations, the VDE (Europe) or nationally recognized regulations relating to the installation and use of electrical equipment and the provisions of the local power supply company.

Pipe penetration (for pressurized gas cylinders in use): The number of pipe penetrations must be limited to the minimum number required and may not exceed three per cylinder. The marked areas of the cabinet roof (inside and outside) and of the side walls (inside only) are to be used for the pipe penetrations.

Purge gas pipeline: Underneath the valve box there are 4 holes for the connection of purge gas lines. Purge gas lines can be connected there. The holes may not be enlarged; smaller purge gas lines must be sealed after they have been fed in. The engaged length of the lines in the rear wall may not exceed 19 mm.

Installing electric cables: The number of cable penetrations must be limited to the minimum number required and may not exceed two per cylinder. The marked areas of the cabinet roof (inside and outside) and of the sidewalls (inside only) are to be used for the cable penetrations. Cable penetrations in the rear wall and doors are not allowed. The maximum allowable diameter per cable is 20 mm and in general, may not exceed the nominal diameter of the cable fed through. Penetrations and holes no longer used must be properly closed off.

Standard roll-in flap: The roll-in flap is unlocked using the latch at the side. The flap operation is a two-handed operation to actuate the latch to prevent uncontrolled falling.

After unlocking the roll-in flap must be folded down to the floor by hand. After unloading or loading, return the roll-in flap back to its starting position by hand (lift up the roll-in flap); the roll-in flap latches automatically in the side latch. The user must ensure that the lock effectively latches into position at the upper initial position.

Cabinet ventilation: DIN EN 14470–2 specifies that the air exchange for a safety cabinet can only be connected to an industrial exhaust air system under the following conditions:

- Designed for use with the cabinet doors closed and locked.
- A minimum of 10 air changes per hour for flammable and oxidizing gases.
- A minimum of 120 air changes per hour for toxic and highly toxic gases.
- The pressure drop within the safety cabinet may not exceed 150 Pa under the above conditions.
- The venting system must cause a partial vacuum in the cabinet.
- The ventilation must be effective over the complete internal height of the cabinet. Extraction can take place through a separate exhaust duct or via a special extractor and must be in operation 24 hours a day (VDI 2051, Item 3).
- The exhaust stream must ensure that small quantities of leaking gases are extracted.
- The exhaust air system must lead to a safe place in outdoors, i.e., discharge into the open air. In the event of a fire, the self-closing supply and exhaust air valves prevent heat from penetrating.
- The venting must be checked by regular visual inspection as well as measurement of the air flow in the cabinet and the pressure drop in the venting pipe.

Note 1: The storage of corrosive gases may have effects on the function of the supply and exhaust air opening shut-off devices and suitable protection or monitoring should be provided.

Note 2: The pressure drop in the cabinet is not the same as the pressure drop at the connection socket of the exhaust pipe. These values can differ from each other significantly especially in case of higher air exchange rates in the cabinet.

Connection to an exhaust air system: The exhaust air and air supply openings are located on the top of the cabinet of the safety cabinet. The cabinet scope of supply includes one exhaust air and one air supply connection NW 0 75, which enables connection to a ventilation system, e.g., DUPERTHAL 2.00.320 for exhaust air monitoring with a ventilator (ATEX-compliant). The venting pipe must be connected to the exhaust air connection socket – left-hand socket viewed from the front. The pipe must be connected to the exhaust air connection socket using a collar, or similar fixing. The air supply can be taken from the room and under normal operating conditions does not require any additional air supply from the laboratory building exterior. After installing the safety cabinet it must be tested to ensure it has been properly connected to the extractor. This includes a visual inspection as well as measurement of the air stream and the pressure drop inside the empty cabinet. The installation of an industrial venting system or connection to an existing exhaust air system may not be part of the DUPERTHAL scope of supply.

The attachment of a ventilator represents an equipment combination. As Zone 2 equipment, the ventilator must fulfill the requirements of the EC Directive 94/9/EC (ATEX) equipment group II, category 3 G.

There is a gate valve under each of the two exhaust air connections of the exhaust and supply air openings. The gate valves are each kept continuously open by a temperature-dependent releasing component (e.g., solder). At a temperature of 70°C the gate valves close the exhaust air and supply air openings. The gate valves and the temperature-releasing component must not be functionally impaired or blocked.

Air ducting system inside the cabinet: The vertically arranged air slits make the venting effective over the entire internal height of the cabinet.

Choice of ventilator: Only ventilators which fulfill the technical specifications defined in the VDMA leaflet 24 169 Part 1 may be used. In non-hazardous areas this means according to VDMA, Indoors: Zone 2 – outdoors: non-hazardous area.

In hazardous areas [3–6] (i.e., potentially explosive atmospheres) all electrical components must comply with nationally recognized regulations and OHS standards (Figures 2.19–2.21).

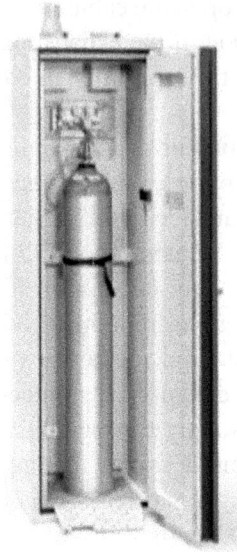

FIGURE 2.19   Duperthal single cylinder cabinet.

FIGURE 2.20   Duperthal two cylinder cabinet.

FIGURE 2.21 Duperthal four-cylinder cabinet.

KEYWORDS

- cryogenic storage
- gas manifold systems
- instrument air plant and equipment
- laboratory cylinder storage cabinets
- laboratory vacuum plant

CHAPTER 3

LABORATORY GAS PIPELINE CONSTRUCTION

3.1 INTRODUCTION

The methods used for the construction of laboratory gas pipelines should be at least equal to the requirements established by national or international standards where available for medical gas pipelines. This would be a recommended minimum requirement as most laboratory gases are considerably cleaner than medical grade gases.

Cleanliness levels for all equipment such as valves, pipelines, and fittings should meet the maximum allowable residual hydrocarbon levels as specified in locally recognized standards and regulations. These levels vary internationally from 2 mg/m^2 in many Scandinavian countries to 30 mg/m^2 in the US and Australia. It is recommended that certification from the manufacturer be sought for the cleanliness level noted by the relevant regulatory authority. CE certification for equipment is available in some countries; however; that is not always the case and additional cleaning may be necessary prior to use. Certification of the cleanliness levels for all equipment being proposed for use should be supplied by the manufacturers as being suitable for use with oxygen and UHP gases as well as being compatible with the individual gas being used.

Tools and equipment should also be in a similar state of cleanliness, it should also be standard practice to ensure the installation staff has an operating manual setting out the use of tools and assembly techniques for this type of work. The procedures for maintaining tools and equipment should have clearly documented and detailed practices in place to have any new or contaminated tools or equipment replaced or cleaned as necessary.

3.2   LABORATORY GAS PIPELINE FABRICATION

3.2.1   COPPER TUBE CONSTRUCTION

3.2.1.1   COPPER TUBE SELECTION

Every country has its own copper tube standard; some are measured by the internal diameter and others by the external diameter. For each outside diameter, there are a number of wall thicknesses available, each is designed for a maximum safe working pressure to meet the system requirements.

The nature of the services included under the label of laboratory gases includes many that are highly flammable, toxic, inert, oxidizing or corrosive and as such special consideration must be given to allowing a wider safety margin than would otherwise be applied. The additional consideration is the availability of oxygen clean tube which will not be available in every diameter and wall thickness provided by the local manufacturers.

The availability of oxygen clean copper tube for medical gas pipeline use is common. The wall thicknesses are generally suitable for pressures well above what is likely to be encountered in laboratory use however there are some situations where that may not apply. Examples of gas pressures that may exceed the safe working pressure for general medical gas grade copper tube are:

a)   High-pressure liquid carbon dioxide used for cooling in GCs.
b)   High-pressure oxygen used to connect to oxygen calorimeters refer to chapter 1, clause 1.4.6 reactors and calorimeters.
c)   Cylinder manifold flexible connections.

Confirmation should be sought to ensure that the copper tube proposed for use is within the safe working pressure of the tube. If there is any doubt, it is necessary to specify a heavier gauge tube along with cleaning specifications for oxygen compatibility and use.

3.2.1.2   COPPER PIPELINE WELDING PROCEDURES

The following is provided with the permission of SAI Global.

On-site brazing of copper-to-copper pipeline and fittings requires a fluxless welding technique with an internal inert gas atmosphere to prevent contamination and oxidation of the internal surfaces of the tube caused by flux residues and oxide scale. The main features of the technique are as follows:

- An inert internal atmosphere of carbon dioxide is used within the pipeline during fluxless brazing. The same procedure is also used during any heating operation involving the laboratory gas pipeline.
- The technique minimizes the fire risk during brazing.
- The absence of flux reduces noxious fumes and flux residue inside the tube.
- The use of carbon dioxide means that monitoring and additional tests and inspections are required during and after the completion of the work on the pipeline.
- Upon completion, the work must be purged with oil-free compressed air or nitrogen with a purity level equal to that of the gas to be used in the pipeline, it must be tested for particulate matter and analyzed to confirm the removal of any residual carbon dioxide.

Silver soldered joints should be made using silver copper phosphorus brazing alloy containing a maximum 15% silver. No flux should be used during the welding procedure. Commercial grade carbon dioxide may be used as the internal inert gas shield to prevent the formation of oxides on the inside of the tubes and fittings. Carbon dioxide should also be used during all tube heating operations, e.g., tube forming and bending where required.

Note: During the purging and welding process the joint to be welded may have an annular gap that can create a venturi effect and a minimal volume of atmospheric air may be drawn into the pipeline, this may cause limited discoloration of the pipe surface.

3.2.1.3   INERT GAS SELECTION

Nitrogen, argon, and carbon dioxide are inert gases and could provide an inert shield for brazing. The use of these gases allows for the creation of an oxygen-depleted atmosphere that may cause asphyxia (lack of oxygen) that can affect an operator should there be no ventilation of the work site. The use of nitrogen or argon as purge gases is not recommended as these gases do not provide an indication of oxygen depletion at the work site and asphyxia may occur without warning.

The information included below is based on the use of carbon dioxide only as the preferred purge gas.

The flushing of pipelines with inert gas during welding processes ensures that silver soldered pipeline joints are constructed using methods to safeguard that the internal surfaces are free of oxides and residues. Experience has shown that this technique is effective, provided the basic principles are followed. The principles include component cleanness, a fluxless brazing method, and correct choice of materials and the use of an inert gas atmosphere during welding.

Since this procedure requires the venting of inert purge gases, it should only be performed outdoors or in naturally ventilated areas. Brazing in non-naturally ventilated areas and in confined spaces requires appropriate safety precautions, including adequate forced ventilation and continuous monitoring of the level of atmospheric carbon dioxide.

Note: National occupational health and safety regulations or other National or Regional Authorities may apply to the use of inert asphyxiant gases on construction sites and similar locations where laboratory gas pipelines are being constructed.

Exposure to 3% or more of carbon dioxide will stimulate breathing. However, higher exposure above 10% will cause unconsciousness.

When using inert gases in confined spaces gas sensors must be used with audible and visual alarm functions.

In non-naturally ventilated areas, the following precautions should be observed:

- Do not exceed the 100 liters/min flush times (see below).
- Ventilate the space by applying forced ventilation.
- Vent the waste carbon dioxide to a safe area by using a connecting flexible hose to the exterior of the work area or by avoiding open-ended tubes that terminate within the workspace.
- On completion, if the pipeline is filled with carbon dioxide after being purged. Flush with instrument grade compressed air and ensure that it vents into the building exterior clear of the work site.
- A risk assessment should be performed to determine whether continuous monitoring is required. Continuous local and personal testing of the concentration of carbon dioxide in the work area must be carried out for the protection of personnel.

Note: If atmospheric carbon dioxide levels above 1.5% are located. The immediate vicinity should be evacuated and the supply source of carbon dioxide isolated and the area evacuated and ventilated. The area should be re-tested prior to re-entry.

On completion of the brazing, the pipeline should again be purged with compressed air to remove residual purge gases and particulate matter.

Pipeline layouts and the procedure for purging with carbon dioxide should be detailed in a work method statement prior to work commencing, with records or project layout drawings maintained of the areas where the work has been completed which should be retained as part of the handover documentation. The work should be planned to reduce the carbon dioxide cylinder movement to a minimum.

Each installation should be planned to ensure:

- Site storage of carbon dioxide cylinders must be provided with secure cylinder storage and restraints.
- The carbon dioxide cylinder regulator should be connected by a flexible hose to the purge/flush valve. The valve outlet should be fitted with a flow metering device which is then connected to the largest diameter pipeline to be brazed. The cylinder valve should be opened slowly until the pressure, prior to the closed isolation valve, reaches the preset pressure of the flow metering device of 400 kPa.
- The complete replacement of the pipeline air by carbon dioxide, without any static 'dead' legs.
- The effective distribution of carbon dioxide during the flushing operation by fitting caps with orifices in the appropriate places.
  Note: Caps create a slight backpressure and therefore force the carbon dioxide through the complete system.
- Carbon dioxide is not applied to sections of the pipeline system that are not currently being brazed.
- Carbon dioxide is not used unnecessarily, e.g., when brazing tube stubs for pressure testing purposes. Tube stubs should extend 300 mm beyond the pipeline to allow removal of the stub's internally oxidized section.
- As a minimum, all laboratory gas pipelines manufactured from copper tube should be constructed as above, the technique applies to all pipelines whether for laboratory gas or vacuum service.

Note: A 30 kg carbon dioxide cylinder contains approximately 17,000 liters of gas and provides approximately 2 hours 45 minutes of flushing or 141 hours and 40 minutes of continuous purging. An equivalent size nitrogen cylinder contains approximately 7,000 liters of gas and may allow for 1 hour ten minutes of flushing or 58 hours 20 minutes of continuous purging.

3.2.1.4   INITIAL PIPELINE FLUSH

The initial pipeline flushing with carbon dioxide should be set at a flow of 100 L/min to ensure that all air is displaced from the pipeline. The duration of the flush depends on the outside diameter of the tube and the length of the assembly. Table 3.1 shows recommended flush times.

TABLE 3.1   Purge Times – Purging With Carbon Dioxide

Pipe size (outside diameter)	Flush duration per 100 meters
15 mm OD	15 seconds
20 mm OD	30 seconds
25 mm OD	45 seconds
32 mm OD	1 minute 20 seconds
40 mm OD	1 minute 45 seconds
50 mm OD	3 minutes
80 mm OD	6 minutes
100 mm OD	12 minutes

After the initial flushing with carbon dioxide, a continuous purge of 2 l/min should be maintained during brazing operations.

Note: If a number of pipelines are purged simultaneously the purge times should be increased accordingly. All laboratory gas pipelines within a localized area should be purged concurrently to ensure adjacent pipelines contain an inert atmosphere to protect against inadvertent heating when adjacent pipelines are being worked on.

When working with large pipeline sections it may be necessary to provide multiple carbon dioxide supply sources and all adjacent lines should be continuously purged. Care should be taken to ensure that the carbon dioxide saturates the pipeline for a minimum of 15 min prior to the commencement of brazing to prevent oxidation may occur inside the tube joint. Selected tube ends should be tested for carbon dioxide concentration using a carbon dioxide analyzer or calibrated test tubes using colored indicators prior to brazing.

3.2.1.5   PIPELINE ASSEMBLY

The pipeline and fittings should be pre-assembled and bracketed in their final position. The brackets should remain loose at this time to allow for tube movement during brazing. Care should be taken to ensure that the ends of the tube and the inside of the fittings are very clean. It is recommended that tube joiners are used to connect straight sections of the tube.

The amount of tube which can conveniently be pre-assembled will depend on the pipe size, location, and complexity of the section to be brazed, a reasonable maximum would be 200 linear meters which would allow control of the carbon dioxide purge gas distribution.

3.2.1.6   BRAZING OPERATIONS

It is important to ensure that the carbon dioxide cylinder used for purging contains sufficient gas for the expected operation. Brazing should be started adjacent to the carbon dioxide source and should be progressed towards the further end of the pipeline. As work progresses, the position of blanking caps may be changed to improve the flow of carbon dioxide into the brazing area. Selection of the location of caps is dependent on the configuration of the pipeline and must suit the section of pipeline being worked on (Figure 3.1).

Note: Care should be taken to protect adjacent laboratory gas pipelines as inadvertent heating of other services could create internal oxidation unless such pipelines also contain an inert gas. It may be necessary to use heat protection devices, e.g., flame guards and fire resistant cloths, to protect the adjacent pipeline, building surfaces or electrical installations from damage.

On completion of the brazing operation, all tube ends must be protected to prevent ingress of dirt or moisture. All pipeline brackets must be completed and pipeline identification labels attached.

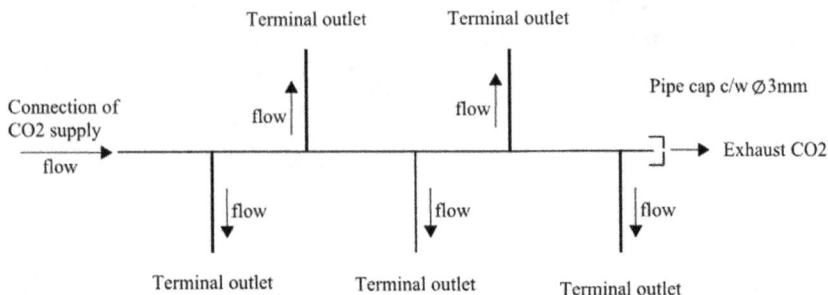

FIGURE 3.1   Sample pipeline diagram indicating suggested blanking cap location.

Note: For carbon dioxide, the compressed carbon dioxide in the full cylinder is in a liquid phase and the pressure cannot be used as an indicator of approaching supply failure. When the liquid has completely evaporated into its gaseous state the gas pressure in the cylinder will fall rapidly. A record of carbon dioxide use should be kept to avoid the possibility of exhausting the cylinder during flushing or brazing operations.

It is important that proper records of carbon dioxide cylinders on the site be maintained. All carbon dioxide cylinders should be accounted for and care taken to ensure that empty and full cylinders are kept separate and they are not confused with laboratory gas cylinders.

Note: Additional information on carbon dioxide may be found in Chapter 5.

3.2.1.7   INSPECTION OF WELDED JOINTS

The following is an extract from AS 2896–2011. © Standards Australia Limited. Copied by James (Jim) Moody with the permission of Standards Australia under License 1703-c120.

On completion of the brazing process, the site engineer may request 1 in 200 or up to a total of five brazed joints to be cut out for examination. Alternatively, there may be an agreement to test joints on a bench at the commencement of the contract to prove the competence of the technician and the effectiveness of a procedure. Removed joints shall be quartered longitudinally, opened, and examined. The tubes and fittings shall be free from oxides and particulate matter (some heat burnishing is acceptable). Owing to the tolerance, the capillary spaces on the pipes and fittings may not have a full penetration of the brazing alloy. Full penetration is not necessary. However, the minimum penetration at any point on the joint shall be more than three times the wall thickness of the pipe and in any case more than 3 mm.

Note: It is recommended that 15% silver content silver solder is used and it should be noted in the specification with advice on the maximum number of connections that may be removed for inspection.

3.2.2 STAINLESS STEEL TUBE

Stainless steel should be equal to the requirements of EN 1027 D4/T3, EN 10216–5, ASTM S213AW/269, EN 1.4401/4404 and AISI 316/316L. It should be oxygen cleaned suitably for use with UHP gases. If available CE Certification from the manufacturer or the independent certified cleaning agent should be provided if not available certification from the manufacturers stating the level of residual non-volatile hydrocarbons that may be found should be provided.

Note: This level of cleaning may not be available in some countries and local pipe manufacturers. Gas suppliers or installation specialists should be consulted regarding sourcing oxygen clean stainless steel tube and the availability of companies who are able to provide suitable cleaning services.

The selection of the type of stainless steel tube provides a number of variables that do not occur in other pipeline materials. The tube is available as seamless or welded seam with a variety of surface finishes. Some are measured by the internal diameter and others by the external diameter. For each outside diameter, there are a number of wall thicknesses available, each is designed for a maximum safe working pressure to meet the system requirements.

The selection of the tube for any project will depend on external restrictions that occur due to the limited numbers of locations where the tube is manufactured. It is necessary to contact the distributors to determine what types, finishes, and wall thickness are available. It may also be necessary to provide a specification for cleaning the tube for laboratory gas use to the minimum required for medical grade oxygen service.

The installation of a stainless steel tube may be carried out using dual olive compression fittings. These are available in a large variety of types, styles, and materials. It is recommended that the selection of materials be made to match the selection of the tubing being used for the main pipeline.

3.2.3   ELECTROPOLISHED STAINLESS STEEL TUBE

Procedures used for the orbital welding of the electropolished tube have not been included in this text due to the variety and complexities of available equipment. For information on the use of orbital welders and associated equipment, it is recommended that the manufacturers of the equipment be contacted.

The stainless steel tube may be supplied as electropolished and cleaned if there is a proven need for this type of tube. This type of tube is manufactured for the electronics industry and electropolishing is a process whereby the tube is internally chromium plated. The tube is available in various surface finishes that are used to prevent the outgassing of surface trapped gases that may contaminate the gases used in the manufacture of wafers and chips.

If it is necessary to provide this type of tubing the system should be orbitally welded using metal gasket face seal couplings for connections to valves and similar controls.

Note: The use of dual olive compression fittings for pipe connections is not acceptable if it is deemed necessary to use electropolished tube and equipment.

It is recommended that the electropolished pipeline is connected to a high vacuum source to evacuate the pipeline prior to connection of the gas supply to ensure extraction of the bulk of any surface trapped gases that may be present.

Valves, regulators, manifolds, and associated pipeline fittings and equipment attached to the pipeline should also be electropolished for similar reasons.

The electropolished tube should be in accordance with SEMI C3–0413 Specification for Gases 2013, SEMI E49–1103 Guide for Subsystem Assembly and Testing Procedures-Stainless Steel Systems, SEMI E49–1104 Guide for High Purity and Ultra High Purity Piping Performance, Subassemblies, and Final Assemblies.

Tubing should be 316L grade stainless steel specified to ASTM A270 internally polished bore of approximately 20 Ra maximum (roughness average in micro-inches). The electropolished stainless steel tube is available in a number of surface finishes (Ra) and the preferred level should be specified.

Standard stainless steel dual olive compression fittings are available in an oxygen cleaned condition that is supplied in variable levels. The recommended oxygen cleanliness level should match that of the pipe selected for use. The internal surfaces of many valves and fittings are not available in an electropolished state and are not considered suitable for service where surface trapped gases may contaminate the gas stream.

3.3   IDENTIFICATION OF PIPELINES

Identifying labels should be fixed to laboratory gas pipelines at all junctions, terminations (e.g., at the rear of outlet fittings, this may be part of the terminal unit), pressure regulation, and measuring points, at the entry and exit of bulkheads and wall penetrations and at intervals of 3 meters throughout their length. The identification color for pipeline labels should be as per national or international guidelines.

3.4   HIGH-PRESSURE VALVES AND REGULATORS

3.4.1   HIGH-PRESSURE POINT VALVES

High-pressure valves for cylinder connection must be gas specific and compatible with the gases being installed. Regulator connections should be as per CGA configuration or as per National Standards where required.

3.4.2   ULTRA HIGH PURITY REGULATORS

Ultra High Purity regulators and manifolds must be provided for all UHP gases. Regulators should be a dual stage with seats and seals manufactured from stainless steel. The regulators should be suitable for the maximum working pressure of the gas type being supplied. This information should be available from the laboratory manager.

The regulator or manifold should be fitted with an integrated safety relief valve or if not available a relief valve should be installed downstream of the regulator prior to the first isolation valve with the connection of an exhaust pipe vented to an external location.

The regulator or manifold should be fitted with a purge valve with a suitable connection for an exhaust pipe to be vented to an external location.

Under no circumstances should any valves or regulators using rubber or synthetic seats and seals be used in UHP laboratory gas pipelines.

3.4.3   TERMINAL OUTLET LOCATIONS

Termination isolation valves should be provided in the locations adjacent to the equipment they are to service or at conveniently located positions in the laboratory, access to equipment or on bench tops where portable equipment will be located.

Note: Fixed labels should be fitted at every terminal location indicating the gas name and supply pressure, similar gases should be identified separately, e.g., HP nitrogen and UHP nitrogen.

3.5   TUBE FITTINGS AND ADAPTORS

3.5.1   DUAL OLIVE COMPRESSION FITTINGS

Dual olive compression fittings deliver a leak-tight, gas-tight seal in an easy-to-install, disassemble, and reassemble form. The robust tube grip with patented, dual-ferrule technology is resistant to vibration fatigue and withstands high pressures and temperature extremes.

3.5.2   METAL GASKET FACE SEAL COUPLINGS

Connections using orbitally welded couplings should be provided where connections to the electropolished tube are used and should be supplied to the same oxygen clean condition as the tube. Certification of cleanliness should be provided similar to that provided for pipeline materials. Connections using a metal gasket sealing provide a metal-to-metal seal which delivers leak-tight service from high vacuum to positive pressure systems. These seals are suitable for dismantling and reassembly which reduces lost time when service or maintenance is required.

Note: The use of non-electropolished fittings and valves negates any advantage that may be obtained by using the electropolished tube.

3.5.3 BALL VALVES

Ball valves using dual olive compression connections are available in a variety of materials. This type of compression fitting is easily available as many types of ball valves are manufactured or configured to suit these fittings.

The ball valve features are:

- suitable for general purpose applications;
- suitable for use where minimal restriction to the gas flow is required;
- available with quarter turn manual, electrical or pneumatic actuation; and
- available to meet a wide range of specifications.

Note 1: Selection of an appropriate ball valve for use in a laboratory gas pipeline may have critical design ramifications especially with respect to the safe operation of the system. Detailed analysis of the proposed use of the valve must be undertaken and a full specification for any valve must be provided.

Note 2: Ball valves are not suitable where sudden opening or closing may cause adiabatic compression. This may generate temperatures that exceed the maximum recommended operating conditions for some gases. Ball valves may also create sudden gas pipeline velocity increases that exceed recommended safe limits recognized by National or Regional Authorities or technical manuals that refer to the safe operating conditions for the gases being proposed for any system design.

3.5.4 CONTROL VALVES FOR FINE FLOW CONTROL AND REGULATION

Dual olive compression connection control valves are commonly used throughout a laboratory gas pipeline. There are many suppliers of these types of valves with a wide selection of connection types.

Note: Needle and regulating valves are the preferred types where gradual activation will prevent adiabatic compression.

3.6　LABORATORY TAPWARE

The following has been provided with the permission of Broen and SGF Laboratories Aust.

3.6.1　COMPOSITE CONTROL STATIONS

The use of system integrated control stations with independent isolation valves and pressure regulation as an alternative to bench or wall mounted isolation is suitable where laboratories with different pressure and flow demands are necessary. This type of controller has the advantage of providing a range of pressures throughout the pipeline by providing a higher system pressure that may be required for specific locations while maintaining a constant pressure at the equipment connected to the system (Figure 3.2).

FIGURE 3.2　Gascon laboratory control.

(**Source:** Gascon Australia. With permission.)

The use of independent control stations is also available with integrated molecular sieves, oxygen traps, and moisture removal for individual connections to equipment. The use of molecular sieves and oxygen traps is specific to the equipment being connected to the pipeline and are only suitable for single connects due to the very low flow rates that they are able to supply, usually measured in c.c. per minute.

3.6.2 WHAT IS 5.0 PURE GAS?

Pure 5.0 gases are typical industrial and technical gases that are 99.999% clean and should be used in applications where impurities must be avoided. Most commonly used pure gases are: argon, helium, carbon dioxide, nitrogen, hydrogen, and methane.

The pure gases are used in innumerable applications in industries such as:

- chemical laboratories;
- pharmaceutical laboratories;
- healthcare pathology laboratories;
- research and analysis laboratories;
- food and beverage testing laboratories; and
- electronics and engineering laboratories.

3.6.3 LABORATORY TAP TYPES AND USES

There are many different types and models of laboratory tapware, the following is based on the BROEN range of laboratory taps (with permission) and includes a selection that is regularly used in laboratory UHP gas pipeline systems. A specification for each type is included for information and includes the recommended gas purity and characteristics of each type.

3.6.3.1 BROEN-CLEANLINE 501 RANGE

BROEN CLEANLINE 501 fittings can be used for gases with purity coefficient up to 5.0 (99.999% clean gases). These fittings can be used when the requirements for gas purity are crucial in laboratories. Also, these fittings can be used for gas chromatography, spectrometry, and for non-corrosive and inactive gases, e.g., argon, helium, nitrogen, hydrogen, methane. They have good pressure control for reliable service. There are also many different mounting possibilities. Each fitting is packaged in a sealed bag to maintain cleanliness. All fittings are leak tested by different pressure methods. One out of one hundred valves is helium leak tested (Figure 3.3).

FIGURE 3.3   BROEN bench mounts single regulator outlet.

Specification

   Max working pressure: P inlet = 21 bar (=305 psi = 2100 kPa = 2100 kN/m^2)

   Max working pressure for single pressure regulation valves: P inlet = 50 bar/360 psi

   Max test pressure (with closed valve): P test = 1.5 x Max P inlet = 32 bar (–464 psi = 3200 kPa = 3200 kN/m^2)

   Outlet pressure: Standard Cleanline 501 fittings are supplied with pressure regulators having the following pressure ranges: P outlet 1 = 0 to 3 bar/44 psi, P outlet 2 = 0 to 10 bar/145 psi

   Pressure regulator for high flow applications: P inlet = 25 bar/360 psi, P outlet = 0 to 8 bar/116 psi

   Nominal flow: 12 Nm^3/h(N2)

   Leak rate: <10 mm^3/s at 6 bar. 1 out of 100 valves is helium leak tested

Inlets*: All fittings have female 1/4" NPT – ANSI B1.20.1 thread connection. Fittings for table/suspended and panel mounting have also male 1/2" NPT – ANSI B1.20.1 thread connection.

*Pressure regulator 02 451.0 x 3 has 2 inlet and 2 outlet connections with female G3/8" – ISO 228/1 thread.

Gas-wetted materials: Body and inner parts – brass. Seals – PVDF, FPM. Manometer – brass. Pressure regulator diaphragm – stainless steel alloy.

Built-in wall mounting fitting with flow and pressure regulation valves, single, and dual-outlet configurations (Figures 3.4 and 3.5)

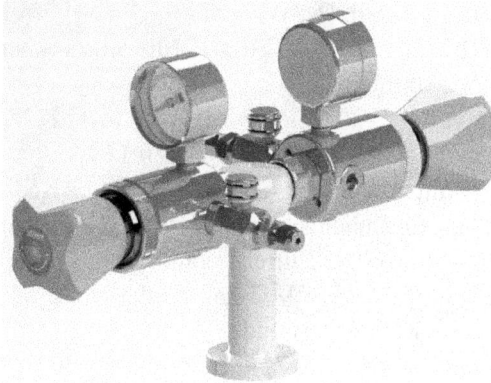

FIGURE 3.4   Broen dual regulator.

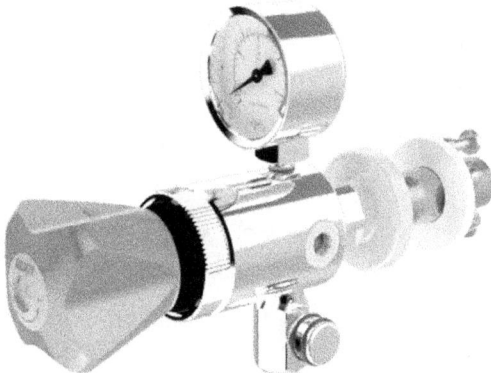

FIGURE 3.5   Broen wall mount single outlet.

Specification

   Max working pressure: P inlet = 21 bar (=305 psi = 2100 kPa = 2100 kN/m²)

   Max working pressure for single pressure regulation valves: P inlet = 50 bar/360 psi

   Max test pressure (with closed valve): P test = 1.5 x Max P inlet = 32 bar (–464 psi = 3200 kPa = 3200 kN/m²)

   Outlet pressure: Standard Cleanline 501 fittings are supplied with pressure regulators having the following pressure ranges: P outlet 1 = 0 to 3 bar/44 psi, P outlet 2 = 0 to 10 bar/145 psi

   Pressure regulator for high flow applications: P inlet = 25 bar/360 psi, P outlet = 0 to 8 bar/116 psi

   Nominal flow: 12 Nm³/h (N2)

   Leak rate: <10 mm³/s at 6 bar. 1 out of 100 valves is helium leak tested

   Leak rate: <5 x 10 mbar x I/s

   Temperature range: –20°C to +50°C/–4°F to +122°F

   Outlets*: Female 1/4" NPT – ANSI B1.20.1

   Inlets*: All fittings have female 1/4" NPT – ANSI B1.20.1 thread connection. Fittings for bench/suspended and panel mounting have also male 1/2" NPT – ANSI B1.20.1 thread connection.

   *Pressure regulator 02 451.0 × 3 has 2 inlet and 2 outlet connections with female G3/8" -ISO 228/1 thread.

   Gas-wetted materials: Body and inner parts – brass. Seals – PVDF, FPM. Manometer – brass. Pressure regulator diaphragm – stainless steel alloy (Figure 3.6).

FIGURE 3.6   Broen Suspended mounting fitting with isolation stop and regulator outlet.

Specification

   Max working pressure: P inlet = 21 bar (=305 psi = 2100 kPa = 2100 kN/m²)

   Max working pressure for single pressure regulation valves: P inlet = 50 bar/360 psi

   Max test pressure (with closed valve): P test = 1.5 x Max P inlet = 32 bar (–464 psi = 3200 kPa = 3200 kN/m²)

   Outlet pressure: Standard Cleanline 501 fittings are supplied with pressure regulators having the following pressure ranges: P outlet 1 = 0 to 3 bar/44 psi, P outlet 2 = 0 to 10 bar/145 psi

   Pressure regulator for high flow applications: P inlet = 25 bar/360 psi, P outlet = 0 to 8 bar/116 psi

   Nominal flow: 12 Nm³/h (N2)

   Leak rate: <10 mm³/s at 6 bar. 1 out of 100 valves is helium leak tested

   Leak rate: <5 x 10 mbar x I/s

   Temperature range: –20°C to +50°C/–4°F to +122°F

   Outlets*: Female 1/4" NPT – ANSI B1.20.1

   Inlets*: All fittings have female 1/4" NPT – ANSI B1.20.1 thread connection. Fittings for table/suspended and panel mounting have also male 1/2" NPT – ANSI B1.20.1 thread connection.

   *Pressure regulator 02 451.0 x 3 has 2 inlet and 2 outlet connections with female G3/8" -ISO 228/1 thread.

   Gas-wetted materials: Body and inner parts – brass. Seals – PVDF, FPM. Manometer – brass.

   Pressure regulator diaphragm – stainless steel alloy.

3.6.3.2   BROEN-LAB CLEANLINE 502 RANGE

The suitable notes for 99.999% clean gases (5.0) are:

- Uniform design and features as for BROEN-LAB™ UniFlex™ fittings;
- Ultrasonically cleaned components;
- Seals are made of FKM and PVDF;
- Metal diaphragm type pressure regulators;
- Good pressure control for reliable service;
- Multiple mounting possibilities;
- Swagelok® tube fittings, secure clean connection on the outlet side;

- All fittings are leak tested by differential pressure method. Leak rate: < 10 mm³/s at 6 bar [CA]; and
- 1 out of 100 valves is helium leak tested. Leak rate: < 5 x 10⁻⁵ mbar x l/s [He].

Double Fittings: BROEN-Lab offers double fittings as a part of Cleanline 502 product range. Use of double fittings contributes to a reduction of installation costs and results in a more efficient utilization of the working space (Figure 3.7).

FIGURE 3.7   BROEN dual outlet configurations.

3.6.3.3   BROEN-LAB CLEANLINE 502 HEADWORKS

BROEN-LAB Cleanline 502 fittings for non-burning 5.0 gases can be delivered with three different types of headworks depending on the requirements set by applications with regards to regulation of media flow:

- high flow;
- needle valve flow regulation; and
- microflow regulation for fine flow control.

3.6.3.4   BROEN- CLEANLINE 601 RANGE

BROEN CLEANLINE 601 fittings can be used for gases with purity coefficient up to 6.0 (99.9999% clean gases). These fittings can be used when the requirements for gas purity are crucial in laboratories. Also, fittings can be used for gas chromatography, spectrometry, and for non-corrosive and inactive gases, e.g., argon, helium, nitrogen, hydrogen, methane. They have good pressure control for reliable service. There are also many different mounting possibilities. Each fitting is packaged in a sealed bag to maintain cleanliness. All fittings are 100% leak tested (Figure 3.8).

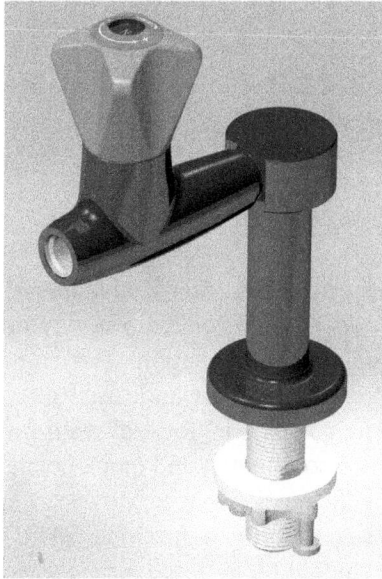

FIGURE 3.8   BROEN bench mounted flow control valve.

Specification

    Max working pressure: P inlet = 21 bar (=305 psi = 2100 kPa = 2100 kN/m^2)

    Max working pressure for single pressure regulation valves: P inlet = 50 bar/360 psi

    Max test pressure (with closed valve): P test = 1.5 x Max P inlet = 32 bar (=464 psi = 3200 kPa = 3200 kN/m^2)

    Outlet pressure: Standard Cleanline 601 fittings are supplied with pressure regulators having following pressure ranges: P outlet 1 = 0 to 3 bar/44 psi, P outlet 2 = 0 to 10 bar/145 psi

    Pressure regulator for high flow applications: P inlet = 25 bar/360 psi, P outlet = 0 to 8 bar/116 psi

    Nominal flow: 12 Nm^3/h (N2)

    Leak rate: <10^{-7} mbar x I/s

    Temperature range: –20°C to +50°C/–4°F to +122°F

    Outlets*: Female 1/4" NPT – ANSI B1.20.1

    Inlets*: All fittings have female 1/4" NPT – ANSI B1.20.1 thread connection. Fittings for table/suspended and panel mounting have also male 1/2" NPT – ANSI B1.20.1 thread connection.

*Pressure regulator 02 451.0 x 3 has 2 inlet and 2 outlet connections with female G3/8" – ISO 228/1 thread.

Gas-wetted materials: Body and inner parts – brass. Seals – PVDF, FPM. Manometer – brass. Pressure regulator diaphragm – stainless steel alloy.

3.6.3.5 BROEN QUICK CONNECT LABORATORY OUTLET TAPS

BROEN quick connect range has many advantages. The products are a flexible and compact design and contain a non-return valve in outlets to protect equipment against contamination. The valves have gas specific keyed systems to protect against an undesired mix of media; 11 media keys are available. The valves are "locked" with no plug inserted. They contain an O-ring which seals outlet and quick connect and is replaceable without disassembling the quick connect. Valves are simple and rational to install. There is also the possibility to install an additional quick connect in the laboratory using a model for exposed piping.

3.6.3.5.1 Bench and Suspended Quick Connect Outlets
(Figure 3.9)

FIGURE 3.9   Broen Bench mounts quick connect outlet.

Specification

   Operating pressure: 1 bar to 10 bar and 14.5 psi to 145 psi
   Purity: 5.0
   Test pressure: 1.5 x operating pressure
   Operating temperature: 0°C to 60°C
   Standard media: Inactive gases (Air, Vac, O_2, N_2, CO_2, H_2, Ar, He)
   Inlet: 1/4" NPT female
   Materials in media contact: Stainless steel, brass and PVDF
   Gaskets: FKM
   External parts: Stainless steel, chemical nickel, plated brass and EPDM
rubber

3.6.3.5.2   Built-in and Wall Mounting Quick Connect Types
(Figures 3.10 and 3.11)

FIGURE 3.10   BROEN built-in wall mount outlet.

FIGURE 3.11   BROEN wall mount outlet.

Specification
   Operating pressure: 1 bar to 10 bar and 14.5 psi to 145 psi
   Purity: 5.0
   Test pressure: 1.5 x operating pressure
   Operating temperature: 0°C to 60°C
   Standard media: Inactive gases (Air, Vac, O_2, N_2, CO_2, H_2, Ar, He)
   Inlet: 1/4" NPT female
   Materials in media contact: Stainless steel, brass and PVDF
   Gaskets: FKM
   External parts: Stainless steel, chemical nickel, plated brass and EPDM
rubber

3.7 GAS SENSING SYSTEM

A gas sensing system should incorporate a central controller that receives
inputs from the various gas sensors and remote inputs throughout the
laboratory and gas storage facility. The controller should provide audible

and visual alarms and controls from local sensors and provide operating controls for remote equipment such as solenoid valves and connections to additional remote controllers. The unit should provide clean terminals for connection to the BMS and engineering and management personnel as necessary.

All components of the gas sensing system should be manufactured for use in the laboratory environment with materials and type of construction selected accordingly. The intrinsically safe design may be a requirement for the system and IP ratings will need to be appropriate for the work setting as will the control wiring between the various parts of the system.

It is recommended that the manufacturers of the gas sensing system proposed to be included in the design of the system to ensure the latest occupational health and safety standards, as well as any local and National or Regional Authorities requirements, are complied with.

3.7.1 GAS CONTROLLERS

The following has been provided with the permission of MSR and Gas Alarm Australia:

The gas controller should be capable of continuous monitoring of the gas sensors and warning of the presence of toxic, flammable, atmospheric oxygen depletion and enrichment levels of the gases piped into the laboratory. The unit should take these inputs and provide audible and visual alarms in locations throughout the laboratory and at locations where the gases are either supplied from cylinders or other sources. The controller should have the facility to connect to a number of locations to inform personnel who use or service and maintain the system including:

- Laboratory personnel in the local area of the safety situation and to take action in accordance with the building management protocols.
- Laboratory personnel in non-local areas who are also interconnected to the gas supply systems.
- Administration staff to activate any procedures (e.g., evacuation) that are required by the facility's OHS safety plan.
- The main fire control board to activate alarms to advise that there may be a safety concern that could be from increased concentrations of flammable, toxic or oxidizing gases.

- Engineering staff to advise that the gases systems have been shut down.
- Any other facility staff that may be affected by a gas failure.
- Connection to the BMS (Building Management System).

The analog inputs should allow for the provision of multiple alarm set points per input and output controls and relays with clean NO/NC contacts for control of solenoid valves for the isolation of the piped gases located in the gas cylinder store.

Alarm output provision should be provided for BMS connections (where available) either analog or digital control of the remote operation of the solenoid or motorized valves on each gas system. Audible and visual alarms should be fitted at each entry of each laboratory that has sensors. Controllers that are able to provide sufficient output connections to operate the audible visual alarms for all gases being monitored should be located on both sides of all entry and exit doors to the laboratories.

The gas controller should incorporate an automatic self-checking protocol to monitor the integrity of analog inputs and the capability to detect short and open circuitry. Where possible the gas controller should incorporate an independent power supply plus backup capable of the operating solenoid and motorized valves and other devices incorporated in the laboratory gas alarm system.

Controllers should provide sufficient analog inputs using 4 to 20 mA or 2 to 10 V DC signals including digital inputs from the Emergency Push Buttons in each laboratory.

Larger systems (beyond 18 to 20+ sensors) can also be installed using an RS485 system where all the sensors are daisy chained over a 2-wire bus. Each sensor is given a unique address (Figure 3.12).

The Gas Controller series MGC–04 can monitor up to 24 analog gas transmitters with 4 to 20 mA signal. Five alarm thresholds are adjustable to each channel. For activating the alarm, up to 20 relays with changeover contacts are available.

The free adjustable parameters and alarm threshold enable very flexible use in the gas measuring technique. Simple and comfortable commissioning is granted by the factory-adjusted parameters. The configuration, parameter settings, and operation are easy to do without programming knowledge. The Gas Controller MGC–04 is equipped with a self-monitoring system and with power supply monitoring. Analog inputs are monitored to detect short-circuit and wire breaks.

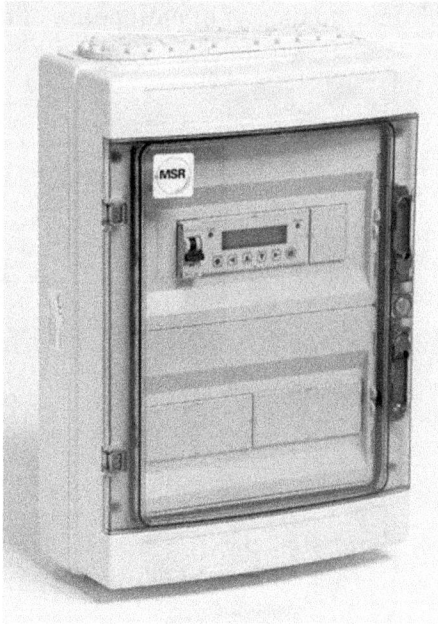

FIGURE 3.12   MSR polyguard MGC 04.
(**Source:** MSR-Gas Alarm Australia. With permission.)

Additionally, the gas controller is available with emergency power supply. This allows for cheaper labor/wiring costs compared to traditional systems; however, this is only suitable for large systems.

The sensor bus is then wired back to the controller; the controller can recognize every sensor as a unique device by its address.

Any number of sensors can be placed within a zone and can be mixed gas types.

Door entrance modules (DEM) for each zone can also be placed at both sides of the entry door of each, also part of the same sensor bus; This allows for each lab owner to see their sensors only on the integrated LCD of the DEM. A DEM also consists of relay outputs and an integrated mute button. If a particular lab has a gas leak only its DEM will activate its relays; the rest of the system will behave as per normal. This further saves the need to wire the A/V (audible/visual) alarms all the way back to the main controller.

Controllers may also have an HLI (high-level interface) for monitoring by the BMS, either mover MODbus RS485 or BACnet/IP. The

HLI will allow real-time monitoring of most points of the control system (Figure 3.13).

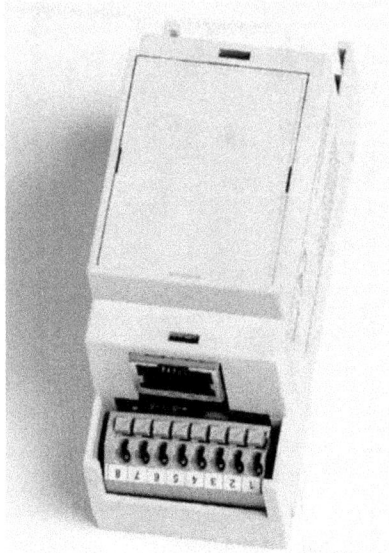

FIGURE 3.13   MSR-BacNet–05 gateway.

(**Source:** MSR-Gas Alarm Australia. With permission.)

The MSR- BacNET–05 gateway is a communication module used to transport values from gas measuring systems or gas sensors to BMS or PLC systems, which are able to be connected via Ethernet to a BACnet/ IP-network.

3.7.2   GAS SENSORS

3.7.2.1   OXYGEN DEPLETION AND ENRICHMENT ALARM

Oxygen depletion sensors should combine an oxygen level transmitter including digital measurement value processing and temperature compensation for the continuous monitoring of the oxygen concentration in ambient air. The unit should have integrated controls including:

• A calibration routine with selective access.

- A standard analog 4–20 mA or 2–10 V DC using an RS–485 or equivalent interface.
- 0–25%VOL measuring range allowing the same sensor to be used for depletion and enrichment applications.
- Output relays including adjustable switching thresholds.
- Dual level oxygen depletion alarm functions for monitoring the laboratory gas atmosphere.
- Oxygen enrichment alarm.
- Automatic synchronization on the multi-sounder audible alarm system.
- Continuously rated.
- Stainless steel fixings.
- Mounting via internal fixing positions or external mounting lugs.
- Duplicate cable terminations (in & out for daisy-chain installations).
- Audible alarms with custom tone configurations and frequencies.

Note: Oxygen depletion alarms are not suitable for the detection of carbon dioxide due to the low detection levels necessary for CO_2 and N_2O. Independent CO_2 and N_2O gas sensors are required if these gases are piped in the local area or if the gas is suspected in the locality. There are other flammable and toxic gases that may also require highly accurate sensing devices (Figure 3.14).

FIGURE 3.14   MSR oxygen gas transmitter ADT–93–1195.
(**Source:** MSR-Gas Alarm Australia. With permission.)

ADT–93–1195 series Oxygen Gas Sensor is a transmitter with digital processing of the measured values and temperature compensation for the continuous monitoring of the ambient air to detect % volume of oxygen (O_2). The transmitter is used within a wide commercial range for detecting flammable gases and vapors.

Features:

- (0) 4–20 mA/(0) 2–10 V analogy signal output, selectable;
- Modular design (Plug-in);
- Simple maintenance;
- IP 65 protected;
- Long life sensor;
- Continuous monitoring.

Optional add-ons:

- RS485 Modbus Interface;
- BACnet/MSTP Interface;
- RS485 DGC (To suit MSR Digital Gas Controllers);
- Relay output;
- LCD Display;
- Heating;
- Integrated Buzzer.

Built for a standard digital signal output, this transmitter is compatible with the Polygard series controllers as well as any other electronic control or automation system.

The sensor unit MC2 houses a module with μ-controller, analog output, and power supply in addition to the electrochemical and cata-lytic Pellistor sensor element including an amplifier. The μ-controller calculates a linear 4–20 mA (or 2–10 V) signal out of the measurement signal and also stores all relevant measured values and data of the sensor element. Calibration is done either by simply replacing the sensor unit or by using the comfortable, integrated calibration routine directly at the system.

Application: The μGard®2 Sensor MC2 is used for the detection of Toxic gases or Combustible gases or for oxygen monitoring wherever a typical 4–20 mA (or 2–10 V) signal is required.

3.7.2.2 FLAMMABLE GAS SENSOR

The flammable gas sensor should incorporate digital processing of the measured values and temperature compensation for the continuous monitoring of the ambient air and detection of combustible gases and vapors. The unit should have integrated controls including (Figure 3.15):

- Calibration routine with selective access;
- The analog output (0) 4–20 mA or (0) 2–10 V DC the ADT–03 and equipped with an RS–485 interfaces for different protocols;
- Dual relays with adjustable switch thresholds are available;
- NRTL performance Tested and Certified Conforms to STD UL 2075 ƒ (UL2075 is only for US applications);
- Digital processing of the measured values including temperature compensation;
- Continuous monitoring;
- Low zero-point drift;
- Long life sensor;
- Housing must be fire-resistant according to UL 94V2 [64];
- Modular design (plug-in);
- Reverse polarity protected, overload protected and short-circuit proof;
- IP65 protected.

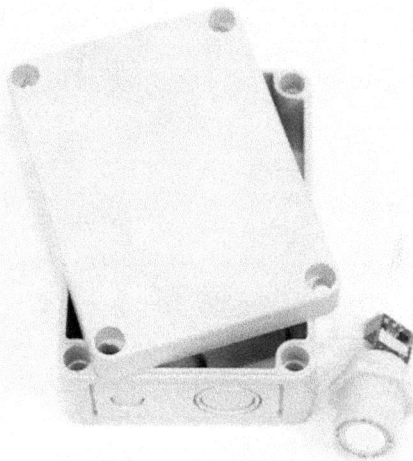

FIGURE 3.15 MSR MC2 sensor transmitter for combustible and toxic gases.
(**Source:** MSR-Gas Alarm Australia. With permission.)

3.7.2.3   CARBON MONOXIDE SENSOR

The Carbon monoxide sensor should incorporate temperature compensation for the continuous monitoring of the ambient air to detect carbon monoxide concentrations. An easy calibration routine with selective access release integrated into the transmitter. Outputs should include analog outputs 4–20 mA or 2–10 V DC, and an RS–485 interface and dual relays with adjustable switching thresholds. The unit should be easily replaceable from local suppliers with service technicians and backup support. The unit should include (Figure 3.16):

- Continuous monitoring;
- Low zero point drift;
- Long life sensor;
- Housing fire-resistant according to UL 94V2 [64];
- Modular plug-in technology;
- Comfortable calibration with selective access release;
- Reverse polarity protected, overload, and short-circuit proof;
- Be at least IP65 protected.

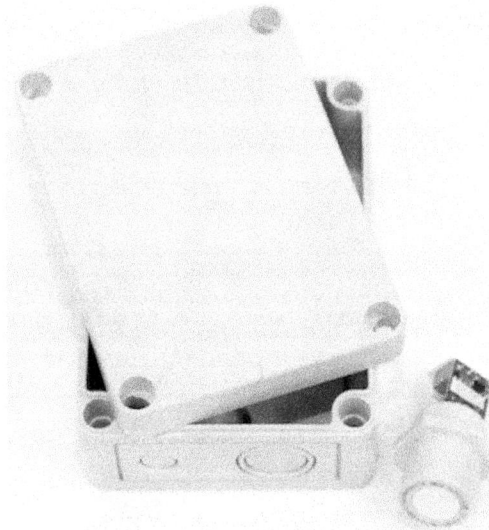

FIGURE 3.16   MSR MC2 sensor transmitter for non-combustible gases.
(**Source:** MSR-Gas Alarm Australia. With permission.)

3.7.2.4 CARBON DIOXIDE SENSOR

The carbon dioxide sensor should incorporate a two-beam infrared sensor for the continuous monitoring of the ambient air to detect carbon dioxide concentrations. It should utilize infrared measuring with integrated temperature and drift compensation and have are calibration interval of 3 years. The sensor should have a standard analog output (0) 4–20 mA or (0) 2–10 V DC and an RS–485 interfaces.

- Two-beam infrared gas sensor (NDIR);
- 0–50,000 PPM (0–5%VOL) measuring range, as the STEL is 30,000 PPM (3%VOL);
- High accuracy, selectivity, and reliability;
- Automatic drift and temperature compensation;
- Good resistance to poisoning;
- Life expectancy > 10 years;
- Comfortable calibration with selective access release;
- Reverse polarity protected, overload, and short-circuit proof;
- IP65 protected;
- Housing fire-resistant according to UL 94V2 [64];
- Modular plug-in technology;
- Approved according to EN 61010–1; ANSI/UL 61010 1; CAN/CSA-C22.2 No.61010–1.

3.7.2.5 AUDIBLE/VISUAL ALARMS

Audible/visual alarms should include a variable volume alarm sounder and beacon with the following inclusions (Figure 3.17):

- Automatic synchronization for a multi-sounder system;
- Continuously rated;
- Stainless steel fixings;
- Mounting via internal BESA compatible fixing positions or via external mounting lugs;
- Duplicate cable terminations capable (in & out for daisy-chain installations);
- Custom tones and frequencies.

FIGURE 3.17   MLD–95A MSR siren and strobe.

(**Source:** MSR-Gas Alarm Australia. With permission.)

MLD–95A is a combination of siren and strobe which can be used for both audio and visual warning/alarming. This unit is typically combined with controllers to provide a complete Gas Alarm detection solution. The unit operates on 24 VDC consuming roughly 250 mA (max). The 3 wire connection provides two modes of independent operation:

- Strobes and Alarm (Flasher) for Lo-Alarm (warning); and
- Siren and Strobe (sounder and flasher) for Hi-Alarm (Final alarm).

3.7.2.6   EMERGENCY CONTROLS

Emergency push buttons should be provided by the electrical trade subcontractor that will allow coordination of a single point alarm activation location. The push buttons should have clean contacts provided for connection to the gas controller.

3.7.2.7 ONGOING CALIBRATION AND SERVICING

A critical part of any gas detection system is its calibration and the replacement of sensor cells upon expiry. Routine calibration should be conducted as per the manufacturer's guidelines; most are annually or bi-annually.

Over time the sensitivity of all gas sensors declines as they age. Calibration will detect or correct this loss of sensitivity.

Depending on use (i.e., how often the sensor detects the gas it is identifying) the harder it works, in turn, reduces the life of the cell and the regularity of replacement may increase.

After a number of calibrations, the sensitivity of the cell cannot be corrected, this will determine when the cell needs to be replaced with a new unit. This period depends on the type of measuring technology incorporated in the cell and how often it detects the gas. For example, a H2 sensor which might detect a leak every day will have a reduced lifespan compared to the same H2 sensor which may detect the gas once a month (Figure 3.18).

FIGURE 3.18 PCE06 Easy Config PC tool.

(**Source:** MSR-Gas Alarm Australia. With permission.)

Product Description:
- PCE06 Easy Config PC Tool is used for the product lines PolyGard ®2/PolyXeta®2.
- PCE06 is a self-sufficient, menu-driven PC tool for comfortable addressing, parameter settings, and calibration of the devices of the product lines PolyXeta®2/PolyGard®2.

Features:
- User-friendly design;
- Installation on PC not required;
- Available for operating systems Windows XP, 8, 10;
- Dongle protection in the adapter; therefore, no problems in changing the PC (turning off virus scanning for USB required);
- No free USB required thanks to internal 4-port USB hub;
- Communication/power supply of the adapter via a cable.

3.8   ENGINEERING-MECHANICAL DESIGN CONSIDERATIONS

3.8.1   ISOLATION VALVE BOXES

Isolation valve boxes should be provided at the main fire exit from each laboratory or work area. Isolation valves must be full bore line size valves compatible for use with each specific laboratory gas and should have been completely dismantled and cleaned suitably for oxygen use by the manufacturer prior to delivery to site. Certification of the cleaning process used should be provided as part of the handover documentation.

3.8.2   PIPEWORK INSTALLATION

The drawings indicate the general location of the pipework; it will be necessary to confirm locations on site. The coordination of laboratory gas pipelines is the responsibility of this subcontractor. All changes in direction of gas services pipework should be achieved by easy bends, full-bore maintained, of radius greater than 5 times the tube diameter, unless otherwise approved.

Ball valve sizes should be sized not less than the size of the pipeline in which the valve is installed. Valves with reduced bores should not be used.

Pipework should be fabricated from the longest possible lengths of tubing in order to minimize joints. Bend tubing wherever possible to minimize fittings; welded joints should not be installed in walls or other inaccessible locations. If no alternatives are available, sections installed in this manner should be pressure tested and witnessed prior to closing in.

Do not install copper in contact with steel, zinc, or other materials likely to generate an electrolytic reaction. Make junctions between dissimilar metals with fittings manufactured in the suitable compatible material. Insulation must be installed between dissimilar metals in these instances.

3.8.3 PIPEWORK SUPPORTS

All pipework should be adequately and securely supported by hangers or supports at proper intervals to prevent sagging.

Install pipework supports in accordance with the Table 3.2, use galvanized two-piece clamps with isolating neoprene cushion or approved equivalent. Each pipeline should be supported and bracketed independently to the channel section; single brackets for multiple pipes may be used.

Brackets should be manufactured from metal; the use of synthetic brackets is not acceptable.

TABLE 3.2 Intervals Between Pipe Supports

Pipe OD in mm	Horizontal distance in meters	Vertical distance in meters
15	1.5	1.8
20	1.8	2.0
25	2.0	2.5
32	2.0	2.5
40	2.5	3.0
50	2.5	3.0
65	2.5	3.5
75	3.0	3.5
100	3.0	4.0

3.9   COMPLETED SYSTEM TESTING

On completion of the installation of the terminal units each section of the piping system should be subjected to a pressure test by the installer in accordance with the following test pressures:

3.9.1   POSITIVE PRESSURE LABORATORY GAS PIPELINES

The pipeline test should be set at 1.5 times the working pressure for the specific gas; the only allowable pressure change in a 4 hour period should be that caused by variations in the ambient temperature existing in the area surrounding the pipeline system.

    Where high-pressure systems are piped it may not be possible to pressurize to 1.5 times this pressure and an acceptable alternative would be to pressure test the specific pipeline at the proposed working pressure using dry nitrogen for a period of 8 hours.

3.9.2   VACUUM PIPELINE

This pipeline should be tested at 140 kPa and the only allowable pressure change in a 4 hour period should be that caused by variations in the ambient temperature around the pipeline system.

    Note: All equipment in direct contact with the pipeline that cannot accept positive pressure must be disconnected from the pipeline and the connections blanked off during pressure testing procedures.

3.9.3   ZONE ISOLATION VALVE AND ASSOCIATED ALARM PANEL TESTS

On completion of pressure tests, each zone isolation valve should be tested for closure. Using the system pressures specified the system should be pressurized with the isolation valve open. The valve should then be closed and the downstream pressure dropped to 70 kPa. There should be no increase in the downstream pressure after 15 min. During this test, each valve should be checked to ensure it isolates the gas in the area defined.

On completion of this test, each Laboratory Gas Alarm panel should be tested for the operation of each alarm function that is controlled from the manifold and at each isolation valve box that is configured for local alarm functions. For each gas the operating pressure should be reduced until such time as the alarm is activated. The activation pressure should be recorded and compared to the recommended pressures used for the system design. Any adjustments necessary should be made and the test repeated. Records taken of these pressures should form part of the commissioning documentation.

3.9.4   PARTICULATE TEST

Each pipeline system should be tested for the presence of particulate matter. This test should be carried out at each terminal outlet during the purity testing of the laboratory gas pipeline.

The flow rate for this test should be 150 L/min for a period of not less than 30 seconds. The test should be carried out using a calibrated orifice using a disposable paper element test filter. The paper element should be inspected after each outlet has been tested. The paper element should be white in color and have a maximum pore size of 5 microns; the paper element holder should have provision to vent waste gases to a safe location.

The recommended test unit is the Millipore/Swinnex 47 mm filter holder with tubing connections for particulate testing of gases using replaceable 5-micron elements. The inlet connection is ¼" BSP that requires a suitable regulating valve or orifice and the exhaust gas is taken from the filter through a push on barb tail to a safe location for venting.

Any discoloration or particulates found on the test paper will fail the test and the pipeline in the local area should be isolated; and the system fully purged until such time as the local area has been retested and passed. Any test papers that show discoloration should be kept for future reference, details of the date, location, and witnesses should be recorded and should form part of the commissioning documentation.

3.9.5   GAS PURITY TEST

As part of the initial commissioning of the laboratory gas system, it is recommended that independent analytical testing for each gas for purity at the furthest terminal unit on each floor or level should be independently verified by an externally accredited and certified laboratory. The purity of the gas should equal the gas suppliers documented literature. Results of this test should form part of the commissioning documentation.

3.10 FOR CONSTRUCTION DRAWINGS

Drawings for use on site should be produced in a 1:100 scale for mains pipeline layouts. Detailed drawings for plant rooms, gas cylinder stores, and operating theatres should be 50:1 and all plant room and wall elevations at 20:1.

Pipeline flow drawings and calculations should be provided indicating the expected pressure variations that may be experienced due to frictional losses in the pipe routes and installed equipment. The maximum calculated pressure drop based on pipeline design flow rates for the laboratory gas pipelines should be less than 5%.

3.11 GENERAL REQUIREMENTS

The installation should be carried out only by experienced laboratory gas installers. In some countries, this is a certification that must be provided prior to the installers beginning work on site.

- Supply all pipe in an oxygen clean state to site sealed with capped ends.
- During construction recap, all open-ended pipes when not immediately under construction and when works are halted for periods in excess of two hours.
- Pipework should be stored above floor level at all times.
- Where threaded joints are employed use only oxygen compatible Teflon tape; certification from the manufacturer should be provided during equipment approvals.
- All equipment, valves, and fittings should be kept in dust-proof bags or containers until ready to be installed in the pipeline.
- All tools and equipment likely to come into contact with the equipment being installed is kept in a clean and oil free condition. Any

contamination found on equipment should be removed from the site and cleaned prior to use.

- Any pipe or equipment that becomes contaminated must be returned to the manufacturer or supplier for replacement or to be re-cleaned prior to return to the site.
- There should be no on-site cleaning of equipment, valves, and fittings. Any contaminated parts should be labeled, isolated, and then removed from the site at the earliest possible time.

3.12   OPERATING AND MAINTENANCE MANUALS

Operating and maintenance manuals should be provided in triplicate and original documents only are used in manuals.

The operating and maintenance manual should include complete details of all equipment provided and included in each gas system; having a separate heading for certification documents, equipment, and pipeline design calculations, reduced size drawings, test results and with headings as follows:

- General Description.
- Laboratory Gas Pipeline No. 1.
- Laboratory Gas Pipeline No. 2.
- .... continuing on for all gas pipelines ...
- Laboratory Compressed Air Plant Details.
- Laboratory Vacuum Plant Details.
- Gas Manifold Details.
- Gas Sensing System.
- Laboratory Gas Alarm System.
- Certification Documentation.
- Equipment Guarantees.
- Design Data.
- Test Results.
- Drawings (A3 or similar size).
- A DVD or CD with the manual and drawings included.
- Copies of full-size drawings should also be provided.

Each heading will include subparagraphs, each with brief descriptions of the equipment included in that system giving manufacturers details, model numbers and types that have been included, where equipment is used for multiple systems (e.g., Alarm Panels) and cross-references should be included to indicate any common usage.

KEYWORDS

- **completed system testing**
- **construction drawings**
- **engineering-mechanical design considerations**
- **gas sensing system**
- **general requirements**
- **high-pressure valves and regulators**
- **identification of pipelines**
- **laboratory gas pipeline fabrication**
- **laboratory tapware**
- **operating and maintenance manuals**
- **tube fittings and adaptors**

CHAPTER 4

LABORATORY GAS PIPELINE DESIGN

4.1 INTRODUCTION

This chapter refers to the site installation of the pipeline and the equipment attached or integrated with it. It does not include the design of plant and supply equipment that is covered in Chapter 2.

The decision to design a laboratory gas system carries with it considerable responsibilities and liabilities. Incorrect design or faulty selection of equipment, sizing of plant, and pipelines or errors in final testing and commissioning may have catastrophic results if incorrectly installed.

Correct design, installation, and testing procedures are of the utmost importance when working with laboratory gas systems as the skills of the installers employed to carry out the works.

The information concerning laboratory gases pipeline installations leaves many designers with limited resources to information and unfortunately, laboratory gases are often considered comparable to medical gases due to lack of other references. This is incorrect, simple examples of the error are the use of flammable or toxic gases used in laboratories which do not exist as medical grade gases. The purity of medical gas is relatively low at 99.5% when compared to laboratory gases which are usually UHP at 99.9995% and the range of incompatibility issues found in medical gases is relatively small compared to gases used in laboratories.

This publication is not meant to be a definitive resource on all of the laboratory gases that are piped. It is a general guide to the problems with the common gases encountered and basic information that will provide guidelines or suggest methods of sourcing the additional information required. It is hoped to assist the design of systems that are safe for the installer to construct and also safe and useful for the laboratory staff who, on completion of the project, may have kilometers of pipelines and many thousands or millions of dollars' worth of specialty gas equipment connected to the system.

4.2   GAS SYSTEM DESIGN

The definition of laboratory gases covers a range of products that include inert, oxidizing, flammable, cryogenic, and toxic gases plus a range of gas mixtures that may be included. There are specific factors that determine how a gas system is designed that may vary according to where the gases are used such as the pressure required the necessary flow and the format it must be supplied in, e.g., cryogenic liquid or gas at ambient temperature. The purity of the gas determines the method employed to transport the gas from its source to the point of use such as in cylinders or as a bulk supply in cryogenic storage vessels.

In this book, a line has been drawn between medical, industrial, and laboratory grade gases each of which has precise working parameters that must be adhered to. The design of industrial and medical grade gas pipelines are not referred to in the text.

In addition to the above, vacuum systems have been addressed for laboratory use that must not be confused with the vacuum or suction used in medical applications each of which has specific requirements. The plant and special design requirements are discussed in Chapter 2.

The generic term "laboratory" includes many instances where gas may be provided for simple services such as Liquefied Petroleum Gas (LPG) and propane that may vary from country to country and may include natural gas used in school laboratories for Bunsen burners. These are not covered in this text as there are plentiful international and national standards as well as government regulations that provide detailed information on these gases.

This book is specifically targeted at the growing range of gases used in research and in clinical laboratories for connection to specialized equipment allowing them to push the barriers of knowledge beyond existing limits. These gases can be the source of great frustration should the gas supply be contaminated or provide incorrect results or worse results that are unable to be reproduced or confirmed, and which may have taken years of work to complete due to pipeline or plant sourced impurities or contamination.

The gases referred to in this text are available in various levels of purity; however, when a gas is discussed the purity level will be a matter of selection decided upon by the end user. The gas specific recommendations made here are for UHP (99.9995% purity) as the recognized acceptable purity level used in laboratories. When dealing with the compatibility issues of UHP gases any recommendations given refer to the specific gas being considered and not as wide-ranging recommendations that refer to a range of gases and purity levels.

Maintaining the gas purity level required will be a matter for the design engineer to resolve to ensure the levels of cleanliness used to construct the pipeline do not alter or change the purity of the gas stream. Any design should be based on providing the gas at the terminal connection at the same purity as supplied from the gas cylinder.

4.2.1 SAFETY INTEGRITY LEVEL, SAFETY INTEGRATED FUNCTIONS, AND SAFETY INSTRUMENTED SYSTEM

Safety integrity level (SIL) [35, 48] can be described as a method of decreasing risk using safe methodologies to actively stipulate a predetermined process to diminish the risk built into the proposed system.

SIL [35, 48] certification or approval is provided on a Scale of 1 to 4 based on the Probability of Failure on Demand (PFD) [36–47]. During continuous operation, the 4 levels are altered to Probability of Failure per Hour (PFH) [34–47]. Certification and testing are carried out in accordance with recognized International or Local Standards or regulations [47, 48]. The scale range begins at Level 1 with Level 4 representing the most demanding requirements for reliability.

A laboratory gas piping system could be described as a Safety Instrumented System (SIS) [62], the inclusion of measuring and sensing equipment that has been certified as being compliant with the relevant SIL [35, 48] level should be documented along with any Risk Assessment [47, 48] that is carried out.

The SIS [62] should be standalone without overriding control functions from external sources or controls that may prevent operation of the SIS [62].

The activation of the SIS [62] alarm functions should allow sufficient time for evacuation of the areas under control within a safe time constraint and to ensure the systems are unable to increase the condition beyond safe OHS standards.

The design of the systems should incorporate equipment with a SIL [35, 48] level that will provide sufficient notice that an unsafe condition exists within a suitable lifespan with ongoing necessary service and maintenance procedures to ensure the SIS [62] operates in accordance with the system design parameters.

The SIS [62] should provide the means to warn personal within the affected area to vacate the area prior to the condition becoming a life-threatening or equipment damaging situation, prevent the condition from worsening and to activate any failsafe functions to close off any gas supplies to the area or facility.

There are needs to be recognition of the safety issues that are encountered in the use of gases that may be flammable, toxic, oxidizing or non-breathable. These properties must be taken into consideration when selecting material and equipment to ensure the safety of the personnel working with or in the vicinity of the systems. There are a considerable number of International Standards and reference texts available that provide guidance on the selection of equipment and the necessary requirements for approvals for use and compliance with national and international regulations.

4.3   LABORATORY FUNCTION

When designing a gas system, the function of the laboratory is the primary factor that will determine what scientific equipment will be installed, which laboratory gases are required and what the system pressures, purity, and flows will be.

The selection of the gases will define the proposed storage requirements for plant and equipment and the areas necessary to safely store the plant and gas cylinders. This, in turn, will be affected by the rules and regulations that will govern where they may be located and what will be required to accommodate these supplies.

The SIS [62] will determine what safety measures will be required such as gas sensing systems and controls for flammable, toxic gases or non-breathable gases including alarm functions and automatic shutdown systems that will be structured to suit the format in which the gases will be provided.

Under many circumstances, it will be difficult to obtain accurate information regarding the equipment being installed and the parameters that the gas system will need to meet. It is impossible to design any gas system unless there is access to this necessary information. It should be available from the architect in the form of a detailed list advising what equipment is

being provided including makes and model numbers for use by the laboratory. For example, a HPLGC or an ICP or an AAS should include the
location where they are to be installed. This should be sufficient information to begin the design process and the manufacturer or supplier must be
contacted for information on the gases required and in what format. Each
of the three instruments noted above require different gas supplies; the
HPLGC may require hydrogen and nitrogen or helium, the ICP uses UHP
argon, and the AAS needs acetylene and nitrous oxide or air.

Note: This is a recommendation only as different suppliers and models
of the equipment may require diverse and even additional gas supplies. It
will be necessary to confirm the various gases required for each item of
equipment and the laboratory may have specific requirements to suit the
work they are carrying out.

These gases should be treated separately from any that are being
provided to fume cupboards, which although in many cases coming from
the same source, may be used for completely different purposes and flows
and pressures will vary accordingly.

The data collected must provide information about what purity levels
for each of these gases are necessary. The accuracy and repeatability of
the sampling they perform depend on the purity of the gas specified by
the manufacturer of the equipment. The equipment manufacturers are the
starting place unless the architect or laboratory manager has prior knowledge that they are able to provide.

The practice of designing every system based on UHP gases will
provide an acceptable level of cleanliness for the majority of laboratories.
However, it may be that this level of purity may be an unnecessary expense
at construction time or the opposite may be the case and it may not be
suitable for some more demanding equipment specifications.

In Chapters 5 and 6 there is a guide to gases and purity levels suitable
for the above-mentioned equipment used in laboratory gas pipeline design.
The architect or laboratory manager should always be the first line of
inquiry to confirm what current equipment demands are, these may change
from supplier to supplier and from year to year. It is always pertinent to
confirm the equipment specifications. A major repair due to incorrect gas
purity or pressure could be a dangerous and or costly exercise.

Gases are supplied in a number of types and pressures each of which
require a different approach in the way they are treated during design
and subsequent installation. As an example nitrogen which is available

in various conditions is classified as Industrial (99.5% pure), High Purity (99.99% pure), Ultra High Purity (99.999% pure), Research Grade (99.9999% pure), OFN (oxygen-free nitrogen), and a large variety of gas mixtures plus as a cryogenic liquid (supplied in a range of purity levels up to 99.999% pure). To add to the uncertainty, various gases are supplied at higher cylinder pressures which may require revised equipment design to handle the increased pressure. Should it be necessary to design the system using high-pressure cylinders, the manifold or equivalent equipment must be specified to meet that pressure range.

Note: The high-pressure cylinders may have different CGA cylinder valve connections.

4.4   BASIC DESIGN PRINCIPLES

The following information must be available prior to beginning to design any laboratory gas system. This list is indicative only as new equipment and procedures are being developed continuously plus up to date national or local regulations will be required to supplement it.

4.4.1   GASES TO BE PROVIDED TO EACH LABORATORY

The first step in designing any system is to determine which gases are to be provided and which equipment they will be connected to. All gases are provided in a range of purities and types and each is used for specific equipment.

It is not uncommon to provide the same gas in different purities to the same laboratory. Gas specific equipment is available that has connections and materials of construction that are designed to suit particular gases. Cylinders have gas specific connections that are in accordance with the Compressed Gas Association (CGA), to prevent incorrect connection of dissimilar gases. There are numerous gases supplied in cylinders of various sizes and pressures some of which are supplied in different states, e.g., carbon dioxide and nitrous oxide are in liquid phase in the cylinder and at considerably lower pressures while nitrogen, argon, air, helium are at 13,500 kPa or higher in the gaseous phase. The carbon dioxide and nitrous oxide cylinders have approximately 17,000 liters of gas and the nitrogen, argon, air, helium has 7,000 liters or more depending on the cylinder pressure being provided.

Note: These pressures and volumes vary from supplier to supplier, always ensure the manifolds, regulators, and equipment are suitable for the gas cylinder supply pressure being provided.

4.4.2   PURITY LEVELS REQUIRED FOR EVERY GAS

The second step is to confirm the purity. The industrial grade gases are unsuitable for high purity applications and the cost ramification of using high purity gases for industrial applications can be significant. The purity of the gas will also impact on the equipment that will be selected during the pipeline design. It may be necessary to use specially selected pipeline materials and valves for UHP gases when an industrial grade gas could use lower quality materials. There is a range of equipment such as manifolds and regulators each of which is manufactured to suit specific purity ranges, selection of the correct equipment will be critical to the performance of the system.

4.4.3   PRESSURES REQUIRED AT EACH ITEM OF EQUIPMENT

The pressure that is required will be determined by the equipment that the gas is being connected to. It may be that a system will be connected to a variety of different items of equipment each requiring different pressures which may require independent local pressure regulation at each terminal outlet to meet these demands. The pressure will impact on the materials of construction. High purity gas regulators in different purity levels may be required is similar pressure control abilities or specific pressure requirements may be necessary. High pressure and high purity applications will require equipment that is designed to meet those conditions.

4.4.4   VOLUME OF GAS REQUIRED AND THE EXPECTED USAGE PERIODS

Flow requirements are variable and depend on what the equipment specification nominates, many items of scientific equipment used regularly such as GCs will use relatively low flows in the mL/min range and phone apps are available to assist with this. Equipment such as Incubators will require intermittent supplies also at low flows; however, AASs demand higher gas flows that may be in use continuously for a number of hours.

The supply of acetylene to an AAS is a special case for consideration. The demands of the AAS are such that the acetylene supply must be at least one cylinder with a minimum capacity of 7,000 liters of gas per AAS. Anything less than that will allow acetone to be drawn from the cylinder along with the gas which will be transported into the AAS. This will contaminate the pipeline and the AAS and require immediate flushing of the system and repairs to equipment to remove the acetone. Flow demands will determine the size of the pipelines as well as the capacity of the plant or cylinder storage necessary. As noted in the acetylene example, insufficient supply may damage equipment or cause failures of sampling and subsequently to sampling procedures.

Selection of supply sources for gases such as nitrogen may demand a cryogenic storage vessel. Compressed air and vacuum will almost always require mechanical plant both of which require suitable, ventilated floor space for installation as well as the coordinated design of the pipelines and plant sizes.

4.4.5   DIVERSITY FACTOR CALCULATIONS

For systems with multiple outlets, many are likely to be used simultaneously, in teaching applications such as Universities and Colleges. The number of independent laboratories could be enough to warrant multiple sources of supply or large central systems that cater for a calculated number of outlets using a pre-ascertained diversity factor. As an example, a single laboratory could expect to have 100% of the outlets being used as all students would be doing a similar procedure concurrently. An industrial or research laboratory could be as low as 10% using the facility due to the number of staff and the nature of their work.

There is only one way to ascertain the diversity factors to be applied, discussions with the laboratory staff to investigate numbers of personnel who would be expected to be using any single laboratory at any one time. If there are multiple laboratories, such as would be found in a University, then the numbers of situations that would be expected could be considerably less. This is applicable to large teaching laboratories, in particular. For small equipment rooms with limited numbers of specialist equipment, it is prudent to allow for 100% use due to their small numbers and the applications that may require hours of continuous use. There is no rule of thumb for these applications, site inspection and discussions with the architect and laboratory manager are the only alternative. Diversity factors will vary according to the type of laboratory as well as the type of equipment being used.

Having collated this information, we move to the next phase of the design; i.e., the selection of the equipment and materials to be used for the installation. The collated data will dictate what can be used and in what locations. However, there will be a variety of options that can be selected. Choosing the right equipment will be difficult as much of the equipment will not be available "off the shelf" due to the exacting requirements the laboratory gas system demands. It may be necessary to source equipment from overseas suppliers such as vacuum jacketed cryogenic pipelines or from local sources that provide close matches that may need additional cleaning or adjustment to meet the demands of the system.

Any movement away from the minimum design requirements will diminish the ability of the system to meet the safety and operational prerequisites of the system.

4.5   EQUIPMENT SELECTION

The equipment used in gas pipelines would ideally be locally sourced from specialist suppliers or agents. Equipment used in laboratory gas pipelines may be gas specific and require an in-depth knowledge of the properties of laboratory gases to ensure manufacturing requirements are able to meet the demands of laboratory gas systems.

Manifolds are constructed to suit the gas, its pressure, type, purity level, specialist controls (e.g., flash arrestors, relief valves, solenoid valves), and flow demands of the systems to which they are to be connected. The same may be said of the manufacturers of laboratory terminal outlets although the designer will be called upon to select the correct configuration to suit the application, e.g., single or multiple benches mounted turret type, valves with dual olive compression fittings (single olive compression fittings should not be used for laboratory gas pipelines) with wall or bench mounted arrangement. Outlet valve types with serrated barb connections used with synthetic tubing are not suitable for use with high and ultra-high purity laboratory gases.

The isolation valves, pipeline control valves, and general engineering equipment can be difficult to obtain due to the limited range of gas compatible apparatus required and the local availability of suitable products. There are specialist ultra-high purity valve manufacturers who provide cleaning services for their equipment and will supply certification of the processes provided. This may not be available from some sources and it may be necessary to provide additional cleaning or preparation from external suppliers.

The cleaning specification for equipment to suit the purity level of the gases being piped determines the standards that necessarily must be met. The use of international and national standards or other recognized specifications should be used where available. Should the supplier or manufacturer be unable to provide proof of cleaning. The responsibility will fall to the design engineer to specify those requirements for the project and who must have an in-depth knowledge of the gas properties, and the suitability of the equipment required, and the cleaning processes that have been used or must be undertaken to meet the specification.

Pipelines, in general, are manufactured from stainless steel or copper. Synthetic materials are rarely used apart from short final connections between the outlet valve and the laboratory equipment. Copper pipe is easily available in "oxygen clean" condition, in most countries. This is provided in accordance with the international or national standards or recognized institutions.

When preparing the pipeline design specification there are a number of points that need to be considered each of which may impact the laboratory gas being piped and the pipeline material, these include:

- There are international suppliers who provide "in-house" cleaning with CE certification. The methodology should be looked into and if the results obtained are equal to or better than those recognized internationally for this purpose they would be acceptable. If there is no documented cleaning process that is recognized by any regulatory body, the design engineer must make an assessment of the product and how it affects the overall pipeline design and the inclusion of a subsidiary cleaning specification may be necessary.

ASTM G88–13 Standard Guide for Designing Systems for Oxygen Service [14] could be consulted for this procedure.

- When using stainless steel, there are a number of guidelines that are accepted internationally such as *ASTM G93–03 Standard Practice for Cleaning Methods and Cleanliness for Materials and Equipment used in Oxygen-Enriched Environments* [15], which could also be used for cleaning of valves and control equipment.

Synthetic pipe and fittings are, in the majority of cases, unsuitable for piped laboratory gases for a number of reasons:

- Specific gases change the properties of materials which may become a fire hazard under specific local conditions. Reference to *ASTM G63–15 Standard Guide for Evaluating Nonmetallic Materials for Oxygen Service* may provide some information on these issues or may provide guidelines for sourcing alternatives.
- The material itself may not be compatible with the gas being piped.
- Synthetic materials may emit esters that could contaminate the purity of the gas.
- The adhesive agents used for joining purposes may be volatile and also emit vapors that could corrupt the gas supply.
- Some synthetic materials are hygroscopic and will allow transport of moisture through pipe walls.
- Some synthetic materials are porous to gases such as Helium.
- The material may not be available in a suitably cleaned condition that would meet the requirements of international or national standards.

Additional factors that must be taken into consideration that may adversely affect the pipeline design apart from materials of construction are the location of pipelines:

- Positioning of gas supply and monitoring alarms should be coordinated with other services alarm systems such as Fire Alarms.
- Location of outlet valves to provide easy accessibility and clearance from other services such as power.
- Area isolation valve location to allow access in an emergency situation preferably on the emergency evacuation pathway.

- Distance of plant and equipment from the end user point for pipeline sizing.

4.5.1 PIPELINE EQUIPMENT SELECTION

The physical properties of each of the gases will dictate the selection of valves and equipment, the major concerns that must be taken into consideration include the following sections.

4.5.1.1 COMPATIBILITY WITH MATERIALS OF CONSTRUCTION

The materials of construction of any valve used for laboratory gases must be 100% compatible. Many gases, such as nitrogen, for example, does not react with most materials; however, some gases have specific requirements:

- Helium should not be used with cast bodied valves due to its small molecular size which allows it to permeate coarse-grained castings. It should have stainless steel valve seals and seats, diaphragms, and working parts in direct contact with the gas should be of metallic construction.
- Carbon dioxide and nitrous oxide have issues with many synthetic sealing materials. They are not aggressive to the materials, however, they may be absorbed into the sealing material which may cause the seal to fail.
- Some gases such as carbon monoxide and hydrogen may have issues when in long-term contact with various types of steel and stainless steel. Structural stress cracking or corrosion cracking or embrittlement may occur if certain conditions prevail and any materials envisaged for inclusion must be investigated and approved prior to selection.

There are a number of materials that are non-flammable when in an air atmosphere. This may change when they come into contact with strong oxidizing gases such as oxygen and nitrous oxide. At high pressures stainless steel is not recommended for use in oxygen-enriched atmospheres without extensive design clarification; the conditions occur, such as adiabatic compression. They may provide sufficient heat to allow the stainless steel to become a fuel source which will burn fiercely. *ASTM G94–05 Standard Guide for Evaluating Metals for Oxygen Service and ASTM G88–13 Standard Guide for Designing Systems for Oxygen Service* [16] should be consulted.

The manufacturer of any equipment that is proposed to be installed in any gas pipeline, flammable, toxic, oxidizing, corrosive or otherwise, must be contacted for their confirmation that the equipment is suitable for the proposed purpose. If there is any doubt about the compatibility of equipment and the condition under which the gases will be used then do not proceed. Consider an alternative that will meet the requirements of available standards and recognized references.

4.5.1.2   FLAMMABILITY

The flammability limits in the air of any gas will determine if national or international standards apply for the particular project, in most countries the storage of oxidizing, toxic, and flammable gases is governed by national Regulations which must be complied with.

The flammability of any gas will also determine the type and construction of the valves selected as will the prevailing local conditions. Control valves, gas sensors, pressure switches and any item that is in direct contact with the main gas stream must be selected with a full understanding of the ramifications of the decision.

4.5.1.3   GAS SENSING SYSTEM

The installation of gas sensing systems are included in any laboratory where there is a laboratory gas pipeline. Depending on the gases being installed, there are a range of specific sensor types that can be installed and interfaced through control systems that can shut down any gas supply in the event of leakage or laboratory fire.

4.5.1.3.1   Gas Sensor Types

Gas sensor types that should be considered include:

- Oxygen sensor – to measure the oxygen level in the laboratory to ensure that the minimum oxygen level does not fall below 19% or increase above 23% (these values may be subject to international or national Regulations and must be confirmed).
- Carbon dioxide sensor – to measure the CO_2 level to ensure it does not increase above 500 ppm.

Note: An oxygen sensor is not sufficiently sensitive to measure when dangerous levels of CO_2 concentration are reached.

- Carbon monoxide sensor – to measure the CO level to ensure it does not increase above the recommended limits and is measured in ppm (this value is subject to international or national Regulations and must be confirmed).

Note: Carbon monoxide is a highly flammable and toxic gas that is easily absorbed by the blood; it is cumulative over time and has an affinity in excess of 200 times stronger than oxygen.

- Flammable gas sensor – to measure the flammable gas level which varies between flammable gases, the gases being piped will determine what levels and types are required.

4.5.1.3.2 Gas Alarm Monitoring System

The sensors should be connected to a central control system that provides a visual and audible indication within the laboratory as well as externally to warn of any critical situation that may exist in an area that may or may not have staff in attendance.

An emergency push button should also be provided adjacent to all of the exit doors to enable anybody leaving due to a system fault to automatically activate the alarms. The push button should also be interconnected to shutoff valves located at the gas supply source.

4.5.1.4 STORAGE TYPE

Many gases have differing physical properties that affect how they are stored and by the cylinders and vessels that are used for storing them. The conditions of storage will affect the volume of gas in the cylinders and in some situations, this will dictate how much can be stored and under what conditions.

This is particularly relevant for liquid gases such as cryogenic liquid nitrogen and argon and oxygen. The dangers that exist in these gases include the temperature at which they are stored and the rate of expansion when the liquid phase is returned to gas. The storage of liquefied gases is frequently nationally regulated and investigations must be undertaken to ensure compliance with those regulations. This may be in the form of a nationally recognized standard or a government regulation.

4.5.1.5   STORAGE PRESSURE

The pipeline pressure applicable for each gas should be regulated to the lowest acceptable pressure for the system. It is always prudent to reduce pressures to the minimum required by the equipment at the source; prior to the pipeline, as a safety and occupational health issue.

4.5.1.6   MAXIMUM VELOCITY

The maximum recommended velocity for any gas should not be reached on any laboratory gas installation; unless the system is operating at critical flows the maximum allowable pressure drop in the pipeline will preclude this. If this velocity is likely to be reached it is outside the scope of this book and reference to the gas suppliers and specialist gas equipment manufacturers should be sought.

The calculation must be done nonetheless to ensure the velocity is within recommended limits to provide a safe operating system. Some points that this takes into consideration are:

- Velocity of particulates that may be introduced into the pipeline and carried in the gas stream and the repercussions of any resultant impact on pipe or valves, e.g., erosion, sparking, seat or seal damage.
- Effect of adiabatic compression in closed systems. Isothermal conditions may not apply to high-velocity systems.
- Pressure loss and system ability to meet design pressure and flow requirements. The following is an extract provided with the permission of ASTM.

ASTM G88–05,5.2.6.2 High Fluid (Gas) Velocities [14] – High fluid velocities increase the kinetic energies of particles entrained in flowing oxygen systems so that they have a higher risk of igniting upon impact. High velocities can occur as a result of reducing pressure across a system component or during a system start-up transient where the pressure is being established through a component or in a pipeline. Components with inherently high internal fluid velocities include pressure regulators, control valves, and flow-limiting orifices. Depending on system configuration, some components can generate high fluid velocities that can be sustained for extended distances downstream. System startups or shutdowns can create transient gas velocities that are often orders of magnitude higher than those experienced during steady-state operation.

Note 2 – The pressure differential that can be tolerated to control high gas velocities is significantly smaller than for control of downstream heat of compression [8] (see Section 5.2.7 for discussion of the heat of compression). Even small pressure differentials across components can generate gas velocities in excess of those recommended for various metals in oxygen service [9, 10].

Equation 4.1 can be used to estimate the downstream gas pressure for a given upstream pressure and maximum downstream gas velocity, assuming an ideal gas and isentropic flow [8]:

$$P_D = \frac{P_T}{\left[\left(\frac{V_D^2}{2g_c KRT_D}\right)+1\right]^K} \tag{4.1}$$

where: P_D = downstream pressure (absolute); P_T = source pressure (absolute); V_D = maximum gas velocity downstream; g_c = dimensional constant (1 kg/N s^2 or 4636 lb in.^{2}/lb$_f$s^2 ft); $K = y/(y-1)$, where y is the ratio of specific heats C_p/C_v (y = 1.4 for O$_2$); R = individual gas constant for O$_2$ (260 N-m/kg °K or 0.333 ft^3 lb$_f$/in.^{2} lb$_m$ °R) [refer, Benning, M. A., Zabrenski, J., S., & Le, N. B., (1988). "The Flammability of Aluminium Alloys and Aluminium Bronzes Measured by Pressurized Oxygen Index," Flammability and Sensitivity of Materials in Oxygen Enriched Atmospheres ASTM STP 986 D. W. Schroll, Ed., American Society for Testing and Materials, Philadelphia, pp. 54–71]; and T_D = temperature downstream (absolute).

4.5.1.7   AUTO IGNITION TEMPERATURE

Many gases will spontaneously ignite when they reach predetermined temperatures. This creates a number of extreme issues especially where pipelines are concerned. The safety of the system is always a concern and it is worth doing the calculations to ensure that the system is designed do not transgress that requirement. *ATSM G88–13* provides a formula that can be used to calculate the resulting temperature caused by adiabatic compression.

The following is an extract provided with the permission of ASTM.

ASTM G88–13, 5.2.7.1 Compression Pressure Ratio [14] – In order to produce temperatures capable of igniting most materials in oxygen environments, a significant compression pressure ratio (P_f/P_i) is required, where the final pressure is significantly higher than the starting pressure (Figure 4.2).

Note 4 – Equation 4.2 shows a formula (the formula shown is based upon isentropic flow relations for an ideal gas) for the theoretical maximum temperature (T_f) that can be developed when pressurizing a gas rapidly from one pressure and temperature to an elevated pressure without heat transfer.

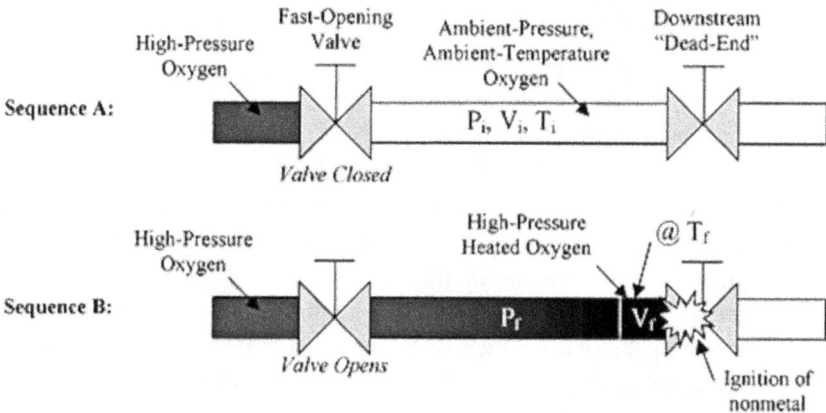

FIGURE 4.2   Example of a compression heating sequence loading to ignition of a nonmetal valve seat.

$$T_f/T_i = [P_f/P_i]^{(n-1)/n}$$ (4.2)

where: T_f= final temperature, abs; T_i = initial temperature, abs; P_f = final pressure, abs; P_i = initial pressure, abs, and

$$n = C_p/C_v = 1.40 \text{ for oxygen} \tag{4.3}$$

where: C_p = specific heat at constant pressure; and C_v = specific heat at constant volume.

Note 1: There are specialized gases used in the electronics industry that spontaneously ignite in air, for example, silane will ignite at a concentration >1.37% in air and special design conditions are necessary for use with these gases. This publication does not address these gases nor offer advice for the design or construction.

Note 2: This book does not provide design information relative to the design and installation of any gases used in the electronics or semiconductor industry.

The compatibility of the materials of construction used for pipelines, valves, and controls and the temperatures at which they will ignite is especially relevant in oxygen-enriched atmospheres. When designing systems where oxygen-enriched atmospheres are present the system designer should read *ASTM G93–03* [15] and *ASTM G94–05* [16] for additional references.

4.6 CRYOGENIC GAS PIPELINES

4.6.1 CRYOGENIC SUPPLY

For the majority of cryogenic gas supplies the storage vessels are leased from the gas supply companies who provide the vessel itself as part of an independent contract with the facility or laboratory. The liquid to gas heat exchanger or vaporizer and the ongoing maintenance and upkeep of the vessel during the lease period would be included as part of that contract. As part of the agreement, they may also provide a gas cylinder backup supply to ensure ongoing supplies are available in the event of a system failure.

The gas suppliers will determine or recommend the size and capacity of the liquid storage vessels; they will be responsible for filling and maintenance. The gas supplier will provide a vessel to suit the expected usage based on their experience and the location of the nearest supply depot. As an example, a vessel located at a remote location would be larger for a similar size facility located in the heart of a major city where delivery can be made daily should that be necessary.

Additional information on cryogenic vessels can be found in Chapter 2.

4.6.2 CRYOGENIC PIPELINES

If the laboratory requires a cryogenic liquid supply to meet the system demand the pipeline is connected to the bottom of the cryogenic vessel and piped directly to the point of use. The pipeline can be manufactured from copper with a polyurethane insulation or using stainless steel using a stainless steel internal pipe with a pressure sealed vacuum jacketed outer casing.

The major distinction between the two types of insulation is the relative performance level. The selection of insulation, no matter which type, will continuously adsorb heat from the atmosphere causing the liquid to convert to its gaseous state within the pipeline. Where the pipelines are installed for continuous service, the use of vacuum jacketed pipelines is recommended to provide the best insulation efficiency for the system and thereby prevent the loss of liquid through evaporation caused by heat transference into the internal pipeline. An example of this type of use would be an MBE that demands continuous cryogenic nitrogen supply for extended periods with a minimal level of previous gas conversion.

Polyurethane insulation may be suitable for systems that are not continuously in use and the cryogenic liquid is not in demand at short notice, examples would be for filling of portable dewars that may only be used irregularly. A relatively short pipeline may also use polyurethane insulation as the limited exposure of the pipeline to ambient temperatures may be acceptable.

Note: The cost difference between a copper pipeline with polyurethane insulation and a stainless steel vacuum jacketed insulation is considerable.

If the system requires a constant supply of liquid it is necessary to provide a means of removing the evaporated cryogenic liquid in its gaseous state from the local application site as well as from the pipeline. This is achieved using a phase separator connected to a high point in the pipeline and as close to the point of use as possible. The phase separator effectively separates the liquid and gas and vents the waste gas to the atmosphere from this high point in the pipeline. This gas is piped to a safe location away from the building preferably at a high level. It should be remembered that this exhaust gas is approximately the same temperature as the liquid and the exhaust pipeline should be insulated using the same material; as the pipeline, failure to provide suitable insulation, will cause the formation of large volumes of ice on the outer casing of the pipeline that will increase in size with damaging results.

There is the added concern that at the termination of this exhaust pipeline at the point of contact with the atmosphere, this exhaust gas will cause a build-up of ice around the exhaust port. To lessen this effect, equipment manufacturers may be able to provide electronic heaters that reduce the build-up and are to be recommended when exhausting these gases.

Stainless steel vacuum jacketed pipelines are generally manufactured in sections which are evacuated and sealed prior to delivery to the laboratory. The vacuum level in these sections is in the vicinity between 10^{-6} kPaA and 10^{-8} kPaA and great care should be taken when assembling the sections and any valves and associated equipment. The outer tube may have a semi-polished finish, nominally 300 grit to assist in reflecting heat from the pipeline (Figure 4.3).

FIGURE 4.3 Vacuum jacketed pipelines during construction.
(**Source:** Cryoquip Australia. With permission.)

The pipeline design will necessarily include changes of direction, isolation valves, solenoid valves, relief valves, and other system controls. These are available from the manufacturers of the stainless steel vacuum jacketed pipelines and are supplied with the same insulation as the pipeline. It is advisable to contact the manufacturers of this equipment to ensure that the valves and equipment required for the pipeline are available at the working pressures that are necessary as well as being suitable for the proposed purposes (Figures 4.4–4.7).

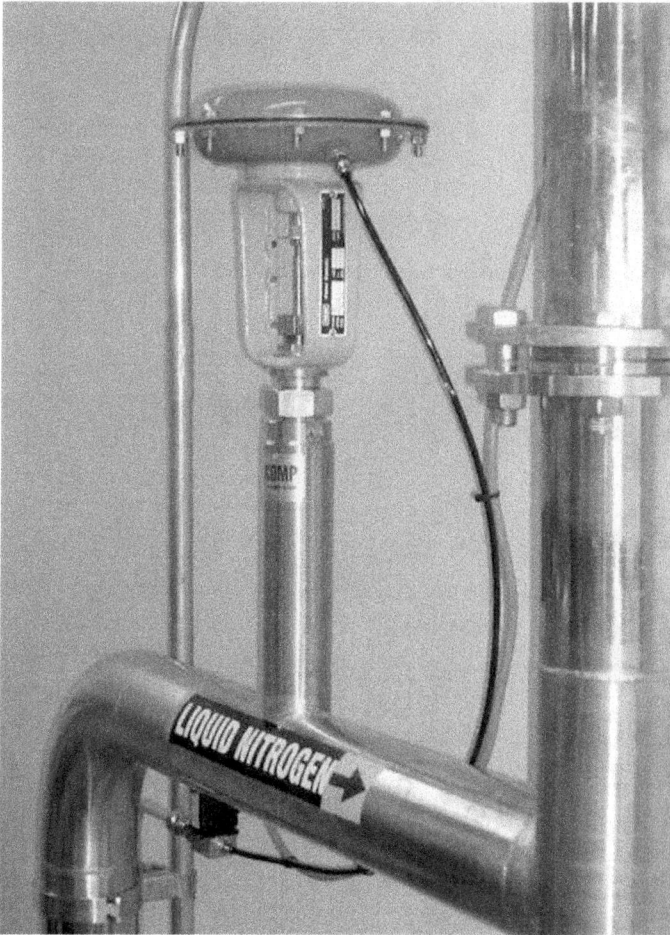

FIGURE 4.4   Vacuum jacketed actuated valve.

(**Source:** Cryoquip Australia. With permission.)

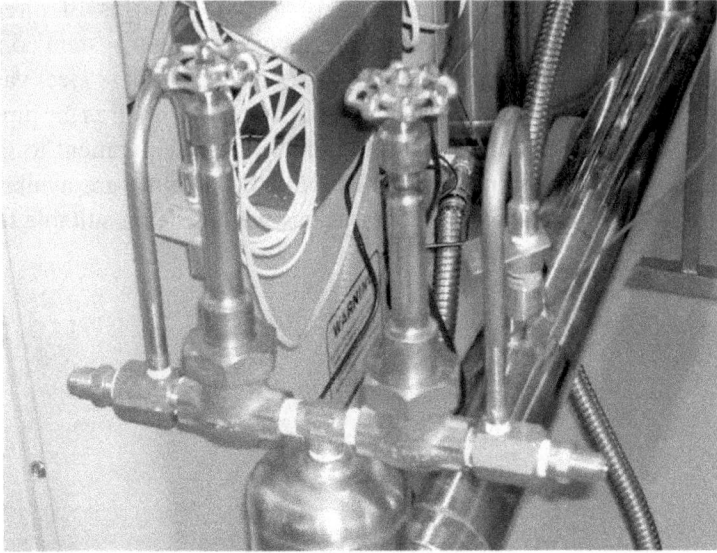

FIGURE 4.5 Extended spindle cryogenic isolation valve.
(**Source:** Cryoquip Australia. With permission.)

FIGURE 4.6 Welded construction vacuum jacketed pipeline.
(**Source:** Cryoquip Australia. With permission.)

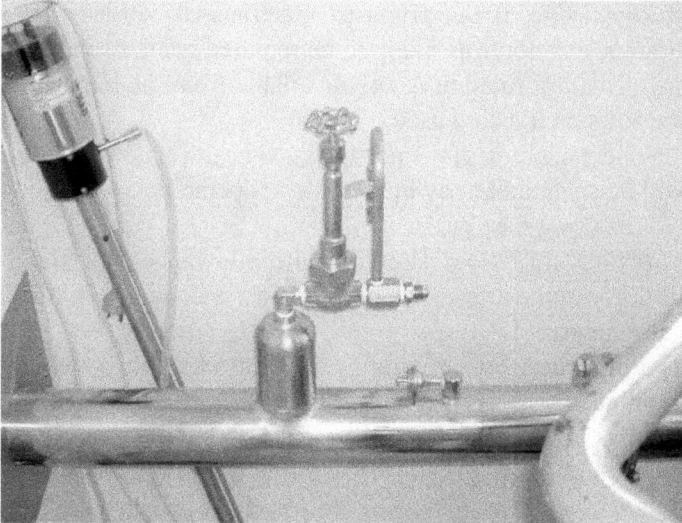

FIGURE 4.7 Welded construction take-off on the vacuum jacketed pipeline with extended spindle isolation valve and pressure relief valve.

(**Source:** Cryoquip Australia. With permission.)

The various pipe sections, changes of direction and any vacuum jacketed valves are either directly welded into the pipeline of connected using a bayonet type coupling that requires considerable space for assembly. These fittings are not flexible and require a suitable distance to allow insertion into the pipeline during assembly. During manufacture each section should be Helium mass spectrometry tested to less than 1×10^{-8} atmosphere cc/sec, to ensure the high vacuum level remains intact and should be regularly inspected to ensure the longevity of the vacuum. Vacuum integrity should be maintained and the vacuum should be guaranteed to be less than 20 microns after 12 months.

Points that need to be taken into consideration when designing a cryogenic pipeline:

- Undertaking the design of a cryogenic system, it should be carried out by a suitably qualified engineer with a complete understanding of the characteristics of the liquids being proposed. Without this knowledge, it is not possible to design a safe working pipeline system.

- A knowledge of the proposed materials of construction, valves, fittings, and controls likely to be required is an essential prerequisite. This information is only available from the manufacturers of the equipment being used.
- An understanding of the occupational health and safety aspects that will be encountered by the installers, operators, and personnel who will be using the system.
- The pipeline layout should make allowances for the installation of rigid sections of pipe, bends, valves, and other control equipment.
- The pipeline layout should make allowance for movement of the pipeline due to temperature variation during use; expansion and contraction of the pipeline for vacuum jacketed systems may not be an issue as this expansion/contraction is allowed for in the internal pipe construction. However, polyurethane insulated systems need to allow for this extension.
- It may be necessary to allow a rise or fall to points of the pipeline to assist in the flow of the liquid and for connection of the phase separators when used. It may also be necessary to ensure that the pipeline is not likely to create gas pockets that will prevent the flow of liquid through the system.
- The phase separator should be designed into the system taking into consideration where it is proposed to be located with sufficient space above the pipeline to accommodate the unit. Phase separators require headspace above the pipeline and as these are proprietary items they cannot be manufactured to fit specific locations.
- It may be necessary to provide a gaseous nitrogen or compressed air supply for control purposes to the phase separator. The manufacturer will be able to provide this information and specific pressure requirements.
- A relief valve should be installed in every section of pipe where possible isolation between valves or terminations of pipelines may occur. Trapped cryogenic liquid will eventually be warmed to room temperature and will expand to many hundreds of times its original volume and within a confined space will increase the pressure to a point where it would rupture any cryogenic pipelines.
- Any relief valves fitted inside the laboratory and connected to the cryogenic system must be exhausted to the building exterior; in the

event of a valve fail during operation it must be remembered the volume of gas that would be vented into the laboratory.

Non-vacuum jacketed valves and controls are readily available from valve suppliers. These are provided with extended handles that can be encased in polyurethane in a similar manner to the pipeline. Standard practice is to encase the pipeline in PVC tubing with supports between the inner and outer pipeline to prevent it from contacting the PVC tube wall. When the pipeline is complete, holes are drilled in the PVC tube and a polyurethane liquid mix is poured into the interstitial space. As the liquid warms it will expand and fill the space and provide a level of insulation suitable for this type of installation. This is not a recommended system for Cryogenic liquids that are connected to phase separators and are expected to be full of liquid continuously as the gas loss due to the inefficiency of the polyurethane insulation caused by heat in leak will be considerable.

Note: Polystyrene is not suitable for use with cryogenic liquids due to its minimum working temperature of approximately –60°C.

4.7   GAS CYLINDER SUPPLY

Cylinder manifolds, compressor, and vacuum plant capacities are calculated from the experience of the design engineer based on a thorough knowledge of the laboratories equipment and usage.

Manifolds for cylinders, designed to provide ongoing supplies, are available taking the same concerns as those for cryogenic systems into consideration. The responsibility for the capacity and design of this equipment rests with the design engineer.

The location of the nearest supply depot will dictate how often the laboratory can be resupplied based on the system demand. For example, a remote site that requires 5 nitrogen cylinders a week and if the local supplier is replenished weekly, then that supplier could be expected to hold 10 cylinders for the weekly delivery which would maintain a regular supply should their supply chain be delayed or interrupted. The manifold would need to be able to hold at least one week's supply; therefore, the manifold would be a 5 x 5 (5 online and 5 in reserve). This would allow a one-week delivery buffer should the need arise. A local supply that is replenished daily could be next day so a 2 x 2 manifold could suffice. The security and financial implications of a larger manifold would also need to be taken into consideration when selecting the right configuration.

4.8   MECHANICAL DESIGN CONSIDERATIONS

There are many variables that affect the stability and operating condition of any gas that is being piped. This paragraph looks at the most commonly found conditions that must be considered when designing any compressed gas or vacuum pipeline system.

Every gas system has specific parameters that dictate the use of materials, types of equipment and the control measures that need to be incorporated into the system to provide a safe working system for the end user. Consideration must be given to all of these factors as many of them will interact with each other that may alter the way they react with other variables in the system.

4.8.1   FLAMMABILITY

Many gases used in laboratories are flammable; these gases have considerably higher flammability levels than the gases encountered in general use such as propane (2.2–9.5% in air) and methane (natural gas) (5–15.4% in air). The use of flammable laboratory gases such as hydrogen (4–75% in air), acetylene (2.5–100% in air), and carbon monoxide (12.5–74.2% in air) have considerably greater flammability ranges and monitoring and management of the systems must be provided with this in mind.

4.8.2   OXIDIZING EFFECT

Oxidizing gases react chemically to oxidize combustible materials. As an example, oxygen increases the flammability of materials and escalates the opportunity for fire or explosion to occur. The reaction may occur spontaneously under ambient conditions or with a minimal additional increase in temperature. Oxidizing gases are considered critical hazards and when designing laboratory gas pipelines consideration must be given to the methodology used for designing the pipeline systems, as well as the storage of these gases to ensure any buildup in enclosed spaces is monitored.

Oxidation can allow combustible materials to ignite spontaneously; many materials that may otherwise not be flammable in air may ignite with no recognizable sources of ignition.

4.8.3   SYSTEM PRESSURE LIMITATIONS

The ideal system design allows for the minimum pipeline pressure to be used that allows for the demands of the equipment it is connected to. In some instances, it may be necessary to provide a higher system pressure to meet localized specific equipment requirements and pressure regulation may need to be provided for areas where lower pressures are required for equipment that is unable to tolerate the higher main system pressure.

The use of flammable and oxidizing gases that have spontaneous ignition points that can be reached under high pressure requires close attention. The occurrence of excessive pressure can be caused by locally generated temperature, adiabatic compression or by incorrect adjustment of pressure regulators.

- The effect of adiabatic compression in pipelines may occur due to the use of incorrect valve types, failure of equipment, and incorrect design of pipelines where changes of pipeline size may inadvertently increase the velocity of the gas stream. The selection of quick opening valves, e.g., ball and solenoid valves, sizing for pipeline systems, are typical conditions where adiabatic compression may occur. Sudden increases in temperature caused by adiabatic compression will cause a change in the conditions within the pipeline such as in the physical characteristics of the materials of construction when operating under changing gas conditions.
- The selection of all equipment used in any laboratory gas pipeline must be suitable for the working pressure the system is designed to accommodate. The system must be provided with sufficient pressure relief safety devices that are capable of exhausting the maximum expected flow rate that would occur if the system pressure regulators fail. This exhaust should be vented at a safe distance away from personnel or if the gas is flammable, oxidizing or toxic it must be vented at a safe location external to the building.

4.8.4   SYSTEM FLOW DEMANDS

The design flow of the pipeline should be based on the gas demand at every terminal outlet to suit the equipment being connected to it. Each location and gas will require a different volume of gas per minute and the use of one flow rate at all outlets throughout any system is not possible.

The only method of calculating the system design flow rate requires a number of steps:

- Discuss the equipment types and models being connected to the pipeline.
- For major items of equipment such as AAS, GCs, ICPs, and similar, additional information is required:
  - Determine the type of usage the pipeline must support and the numbers of personnel who will be using the system.
  - Contact the equipment supplier for major items of equipment to determine the maximum and minimum pressures and flow rate that their equipment needs to operate.
  - Confirm the gas purity that the equipment requires to operate correctly.
  - Discuss with the laboratory manager and the equipment user what if any special requirements are recommended for this equipment such as molecular sieves, oxygen traps, and other locally installed devices.
- Discuss with the laboratory manager and equipment operator if possible the expected operating time for each item of equipment, this will be particularly important when determining any diversity factor that can be applied when calculating the system design flow rate.
- For equipment that has a simple intermittent use such as an oxygen calorimeter, aerator, rotary evaporator, pneumatic operators or similar uses, each of which has vastly different flow and pressure demands. It is imperative that the exact parameters are determined as without them it is not possible to size a supply source or pipeline.
- Flow required from each outlet can be affected by pipeline variables including:
  - Orifice diameter of valves or regulators will create temperature and velocity changes, if the pressure varies due to high flow demands caused by undersized pipelines.

- The terminal outlet distance from main supply point will necessitate correct pipe diameter design to ensure minimal pressure loss at maximum design flow.
- Type of terminal outlet will be determined by the pressure required and usage.
- Mechanical connections may cause restrictions at the terminal outlet.
- Push on hose fittings are only suitable for very low pressure.
- Needle, diaphragm, ball or plug terminal outlets have specific uses each of which determines how the end user may access the gas.

4.8.5   VELOCITY LIMITATIONS

There are maximum recommended velocities for gases that must be strictly adhered to. It would be unlikely to design a pipeline for laboratory use where critical velocities would be encountered due to the pressure loss that would create however it must be taken into consideration using calculations on the expected velocity. The additional crisis that high-velocity gas streams have is the ability to move loose particulates or contaminants within the pipe that may impact on valve seats and seals within the pipeline causing fires or equipment failure and subsequent system and laboratory damage.

4.8.6   COMPATIBILITY WITH THE GAS

Compatibility of the pipeline material with the process that is being undertaken is particularly relevant in laboratory gas pipelines. Some gases have incompatibility issues such as acetylene and copper; and the entry of some VOC's or other contaminants into vacuum systems can cause damage to pipelines, valves, and filtration.

Compatibility of the various gases with the proposed pipeline materials and associated controls that may be incorporated into the system including:

- Valve seats and seals.
- Connection and jointing compounds or materials such as thread tapes and gaskets.

- Cleaning requirements of inline equipment, if necessary, including all lubricants found within the valves and equipment, solvents used during cleaning of components and final removal and flushing on completion.
- It may be necessary to design an oxygen system using *ASTM G93* [15] and *G94* [16] to provide system compatibility of the various materials used to provide a safe working environment.

Compatibility of gas under the conditions in which it will be piped may create additional variables as some gases are compatible with a limited range of materials at low pressures. However, they may have issues at elevated pressures and temperatures.

It may be necessary to provide additional services to the gas stream prior to final connection to equipment (e.g., high-level filtration, molecular sieves or oxygen traps). These are all gas and laboratory specific and would normally be provided by the laboratory or facility staff after completion and handover of the pipeline.

4.8.7 CALCULATION OF PIPELINE SIZES

The common formula used for pipeline flow calculation for non-Newtonian fluids such as compressible liquids is the Darcy Weisbach equation [21–23]. When using this formula there are a number of factors that determine how the flow of gas will behave in the pipeline, these include:

- Pipe roughness factor usually referred to as the Darcy Friction Factor [21].
- The Reynolds number [21–23] calculated for the type of flow commonly encountered in laboratory gas and vacuum pipelines is turbulent, i.e., having a Reynolds number in excess of 2,200.
- The pipe diameter.
- The length of the pipe.
- The number of changes of direction.
- The number and types of valves or restrictions.
- The temperature of the gas.
- The velocity of the gas.
- The density of the gas.
- The pressure of the gas.

The flow of compressible liquids was determined by Reynolds as falling into three ranges: turbulent, intermediate, and laminar.

Reynolds Number [21, 23] references are:

$R > 2,200$ = Turbulent
$R > 1,200$ and $R < 2,200$ = Intermediate
$R < 1,200$ = Laminar

These flow types are calculated and used when working with the Darcy Weisbach Formula and it is common to encounter high Reynolds numbers in excess of 20,000 when calculating gas and vacuum pipe sizes in the pressure ranges used by laboratories. The overriding factor when using this formula is the calculated pressure loss for any given pipe section, the Reynolds number is often very high; however, if the pressure loss is within the limits that the pipeline design can accommodate and the velocity does not exceed recommended values it is not a concern.

4.8.8 GAS PURITY

- Information on gas purity levels used for laboratory gas pipelines is provided in Chapters 5 and 6.
- Purity level of the gas being piped will demand different design criteria for each specific purity level being included. Designs for industrial grade gases would be quite different from the requirements of an Ultra High Purity gas system and would use different equipment and methods of construction.
- The materials of construction for the pipeline, valves, and all equipment plus the construction methods must ensure that the purity of the gas is not altered during its passage in the pipeline between the gas cylinder and the terminal outlet. Testing procedures on the testing and commissioning must incorporate confirmation that the purity has not been compromised prior to handover of the system.

4.8.9 LABORATORY EQUIPMENT ADDITIONAL REQUIREMENTS

If local treatment of the gas stream is required at the terminal outlet for each gas, there are a number of items of equipment that may need to be included in the design of the pipeline. These items include such as:

- Molecular Sieves.
  Note: Molecular sieves are available in a variety of process specific types and cannot be specified without information from the laboratory manager.
- Oxygen traps reduce the oxygen levels to below 15 PPB at flows up to 3 L/min.
- Moisture filtration designed to operate at a single terminal unit and may be incorporated with the oxygen trap.
- Hydrocarbon traps that reduce organic hydrocarbons to less than 0.1 PPM.
- Exhaust of combusted gas waste or exhaust gases treated or removed after the process or procedure has been completed – is there a need for an exhaust system which is particularly important when AASs are in use?

4.8.10   SAFETY ISSUES

- Any Occupational Health and Safety (OHS) issues that need to be addressed for safe working conditions to be provided for the end users, e.g., toxic, flammable gas monitoring, fire safety provisions, materials handling provisions.
- What do the national Regulations or similar local rules and regulations require, and is certification necessary?
- What international standards or similarly recognized authorities are applicable to the installation if no national standard is available?
- What kind of storage facility will need to be constructed including delivery access for cryogenic liquids or cylinders and provision for cylinders in accordance with national or local building regulations and standards?

After considering all of the above, a system design report should be prepared that indicates the formulas used. MSDS's design factors that have been referenced, international, and national authorities that have been accessed and used and any relevant reference documentation that has impacted on the design.

Pipeline designs should also be provided to the client indicating flows, pressure drops and velocities that will be encountered and diversity factors that have been used in the calculation of the design. For gases that have these constraints, proof should be included to substantiate that the maximum recommended levels are not exceeded.

There are references to information included in this book which should be supplemented with information that can be found with research in University or similar facilities, libraries, and databases if necessary to clarify any areas that cannot be resolved from the information provided.

4.8.11 OTHER DESIGN CONSIDERATIONS

Gas pressure is one of the major variables that will impact on the design of any gas pipeline. As pressure increases so do other variables such as the materials of construction of the pipeline, the wall thickness of the pipe being used and the frictional losses that are encountered. That is simple engineering principles; however, when pressure increases other variables come into play that is rarely taken into consideration. Some gases, such as carbon dioxide, change state as the pressure increases from gas to liquid. In its liquid phase at normal cylinder pressure (non-refrigerated), if the temperature should increase above 30 degrees C it will revert to its gaseous phase with associated increases in pressure. If the opposite state is looked into and the carbon dioxide is in its refrigerated liquid state and the pressure falls sharply (which may be due to increased flow), it will form Dry Ice within the vessel and pipeline. Each of these concerns must be taken into consideration when designing the system to suit the circumstances. This gas is available in its various phases from the gas suppliers and each will need special design considerations.

Other examples include high-pressure carbon monoxide (above 5,000 kPa) and acetylene under all conditions. These two gases (among others) will both fall outside the general rules of gas installations, one due to the effect the gas has when piped at elevated pressures and the other due to its incompatibility with a variety of commonly used materials and its own inherent instability.

If we look at the carbon monoxide, this gas, at pressures above 5,000 kPa, may materially affect the structure of a variety of substances. Most designers immediately think that the ideal pipeline material is stainless steel with a suitable wall thickness to accommodate the elevated pressure. This would be incorrect. Carbon monoxide has known physical effects (refer to *Handbook of Purified Gases* by Schon, H. [32]) that over time will degrade some types of stainless steel. A second concern about the installation of this gas at this pressure would be the temperature that could be created under adiabatic compression. At higher pressures, the sudden increase in temperature could exceed the auto-ignition temperature. This property must be taken into consideration for all gases.

When acetylene is used, we are confronted with a totally different set of circumstances. Acetylene gas cannot be installed in a copper pipe. When acetylene comes into contact with pure unalloyed copper in specific instances, it creates acetylides; these are unstable and likely to spontaneously ignite with explosive force. This gas has additional limiting parameters that prevent it from being piped at pressures that exceed 100 kPa as the gas itself becomes unstable and may violently ignite.

Having taken two examples from the large variety of gases that are likely to be piped, it shows how easy it would be for an inexperienced designer or installer to construct a piping system that could prove exceedingly dangerous.

An important gas that should be mentioned here is oxygen; the number of pipelines that are installed throughout the world in hospitals, laboratories, and industrial premises would be numbered in the hundreds of thousands. Oxygen has a number of properties that also create circumstances that are likely to lead to hazardous situations.

The most commonly encountered is in the presence of hydrocarbons within an enriched oxygen atmosphere. To prevent this or mitigate the dangers involved, experienced designers and installers must be employed for both the design and construction of the pipeline.

There are, however, other less well-known conditions such as maximum velocities, compatibility issues with some materials used in locations such as valve seats and seals, compatibility with materials at varying pressures, temperature, and velocity changes through orifices and a number of other commonly encountered installation conditions.

There is a significant amount of engineering information available from sources such as the ASTM G-series documents, gas manufacturers' websites, and MSDSs that are particularly relevant to oxygen (and all other gases encountered in laboratory gas pipelines), and anybody contemplating pipeline design should utilize these resources and become thoroughly conversant with the gas and, indeed, with any gas that they propose to provide pipeline designs for.

4.9   VACUUM SPECIFIC REQUIREMENTS

4.9.1   VACUUM PIPELINE CALCULATIONS

When calculating vacuum pipeline pressure and flows the pressure range is determined as a rough vacuum as classified by Knudsen [21, 59].

Rough vacuum is considered to be in the range of 1,000 mbar to 1 mbar. For general use, piped laboratory vacuum systems are all in the range of Rough vacuum.

There are systems that require vacuum levels to evaporate some VOCs and require a pressure range between –80 and –99 kPa (also considered to be a rough vacuum). These systems require specific vacuum pumps types for this purpose and due to the pressures, it is recommended that pumps are specifically selected for systems requiring pressures in the range below –95 kPa. As an example, it may be feasible to use a Claw type pump to evacuate the system to –95 kPa to reduce the pressure in the rough vacuum range.

If the system is required to operate at –99 kPa then the selection will require the equivalent of an oil-sealed rotary vane or a dry running screw type, if the service has aggressive chemicals and vapors the dry running screw type would be an alternative.

When working with pressures in the vicinity of –99 kPa it is recommended that advice is sought from the vacuum pump manufacturers.

4.9.2   CALCULATION OF VACUUM FLOWS AT VARIOUS PRESSURES

Laboratory vacuum pumps must be able to provide a wide range of operating pressures that have far more demanding conditions than a medical vacuum system and specific plant requirements are necessary to meet these.

The selection of a vacuum pump to suit an application must be investigated. The use of a diffusion pump that will operate at high vacuum levels would not be suitable for rough vacuum and is rarely if ever, used for laboratory pipeline systems due to the pressure range that a laboratory vacuum system will encounter. The differences in construction and efficiency will indicate what pumps to use that will meet the manufacturers' suggested operating parameters. The following gives an example of pump types and the disadvantages of flawed selection.

4.9.2.1   EXAMPLES OF CALCULATIONS FOR A SINGLE LABORATORY AT VARYING PRESSURES

The following example considers a laboratory with 20 workstations in a university student teaching facility where it could be expected that a class of students would be carrying out similar experiments concurrently. It assumes that the equipment being used at each station is similar and each has a flow rate of 40 liters per minute of free air entering the system. The three examples are based on an experiment that proposes to use different VOCs. In each example, we use a different evaporation pressure, i.e., –60 kPa, –90 kPa, and –99 kPa. The purpose is to indicate the increase in the capacity of the vacuum plant necessary for each VOC and why it is imperative that this information must be taken into consideration prior to designing the system.

A laboratory that has 20 terminal workstations with a single outlet per user that would be used in a teaching facility expecting to use rotary evaporators and can be expected to use approximately 40 liters per minute per terminal unit on start-up and end pressures of –60 kPa, –90 kPa, and –99 kPa could be used as a guideline. As the students may all be operating concurrently the flow rate could be calculated using a 100% usage rate for the system as follows:

Note: Calculations below are approximate only and use 101 kPa as the approximate representation of atmospheric pressure.

Boyles Law, P1 x V1= P2 x V2, where:

P1 is initial pressure, abs;
V1 is initial volume;
V2 is final volume; and
P2 is final pressure, abs.

Example 1: evacuation pressure required: –60 kPa

- *P1 = 101 kPaA*
- *V1 = 20 (terminal units) x 40 (L/min)*
- *P2 = 41 kPaA*
- *V2 = Vacuum Pump flow rate*
- Calculation:
    - *V2 = P1 x V1/P2*
    - *V2 = 101 x 800/41*
    - *V2 = 1,970 L/min (118 cubic me/h)*

Example 2: evacuation pressure required: −90 kPa

- *P1 = 101 kPaA*
- *V1 = 20 (terminal units) x 40 (L/min)*
- *P2 = 11 kPaA*
- *V2 = Vacuum Pump flow rate*
- Calculation:
  - *V2 = P1 x V1/P2*
  - *V2 = 101 x 800/11*
  - *V2 = 7,345 L/min (440 cubic me/h)*

Example 3: evacuation pressure required: −99 kPa

- *P1 x V1= P2 x V2*
- *P1 = 101 kPaA*
- *V1 = 20 (terminal units) x 40 (L/min)*
- *P2 = 2 kPaA*
- *V2 = Vacuum Pump flow rate*
- Calculation:
  - *V2 = P1 x V1/P2*
  - *V2 = 101 x 800/2*
  - *V2 = 40,400 L/min (2,424 cubic me/h)*

This is an extreme variation; however, it is an example of sizing vacuum pumps and the correct selection of suitable pump types each of which would exhaust 800 liters per minute of free air for various pipeline pressures:

- At −60 kPa the pump capacity would be a rarefied airflow rate of *1,970 l/min (118 cubic me/h)*.
- At −90 kPa the pump capacity would be a rarefied airflow rate of *7,345 l/min (440 cubic me/h)*.
- At −99 kPa the pump capacity would be a rarefied airflow rate of *40,400 l/min (2,424 cubic me/h)*.

As can be seen from the above, the availability of rough vacuum pumps at a rarefied airflow rate of 118 cu/me/h is relatively simple and can be accommodated by most pump types. However, a pump that is able to operate at −99 kPa requires a substantially larger pump of a different type. A vacuum pump with a rarefied airflow rate of 2,424 cu/me/h at −99 kPa is a specialized pump type and the selection becomes a limited as well as expensive exercise.

In the above instance, the ideal design for each system would have to allow for the design pressures required. Suggested options would be:

- At –60 kPa all pump types would be able to provide this flow rate and pressure, however, the best alternative may be a dry running claw or vane type pump using a VSD drive.
- At minus, 90 kPa, not all pump types would be able to provide this flow rate and pressure and the best alternative may be oil flooded vane type pumps with a stop-start control.
- At –99 kPa very few pump types would be able to provide this flow rate and pressure and the best alternative may be a dry running screw type pump with stop-start control or an independent under bench mounted high vacuum pump located at each workstation.

4.9.3 VACUUM EQUIPMENT SELECTION

When selecting plant it is necessary to take into consideration the compatibility issues dealt with previously plus the effects that the pipeline construction may have on the vacuum pressures. The fluctuations of pressure and flow affect the overall performance of the system and may inadvertently cause detrimental results in remote locations, some of these include:

- The maximum pressure drop at any terminal outlet may vary when pipe sizes are inadequate, this can result in flow reversals throughout the pipeline and this must be taken into consideration when deciding what minimum operating pressure will be acceptable and what the results of any pressure variation in the system may be.
- Flow reversal in vacuum pipelines could allow inadvertent mixing of samples between laboratories with damaging results.

4.10 CLIENT EXPECTATIONS

The client can reasonably expect that the purity of the gas will not alter during transport from the gas supplier to the time they connect to the gas outlet on the pipeline. This will require compatibility with the materials of construction of pipelines and controls and prevention of contamination from within the system. There are a number of compatibility issues that are noted above however there are local and construction concerns that need to be considered:

- The on-site storage of supplies prior to site installation must provide suitable conditions that will prevent accumulation of moisture or particulates during site storage.
- Site work and equipment configurations may include electrical work that includes solenoid valves, gas sensing systems, pressure switch alarm controls for monitoring the gas supplies and storage. During the gas supply storage prior to use, protection from the weather and any site climatic conditions that may occur is necessary.

The client will assume that the pressure and flow will not fluctuate during operation, this is easily covered during the design phase, however, input from the architect or laboratory manager is necessary. Pressure and flow fluctuation may be caused by:

- Inadequate pipe sizing that restricts flow during periods of high demand.
- Insufficient capacity at supply source.
- Pressure loss through undersized pipeline filters if fitted, such as bench mounted oxygen traps and molecular sieves.
- Restrictions caused by control valves with reduced size orifices.
- Single-stage pressure regulators that increase downstream pressure as the upstream pressure degrades and vice versa.

The client expects that any special properties that pertain to the gases have been taken into consideration when designing the system such as:

- Maximum safe pipeline pressure settings and maximum output from cylinders to provide clean free gas supply.
- Pressure regulators have metal seat and seals to prevent diffusion through the seats and seals.
- Compatibility of the gas with seats and seals, e.g., carbon dioxide and nitrous oxide are incompatible with some synthetic seats and seals.
- Compatibility of the gas within the pipeline is ensured, e.g., acety-lene is incompatible with copper.

- Suitability of the regulators, valves, and terminal outlets with the flows and pressures for each gas, e.g., ball valves are unsuitable for use with liquid gases.
- Gas properties will not change as pressure or flow changes, e.g., some cryogenic liquids alter state to the gaseous phase if the pipeline insulation is inadequate.

4.11 COMPLETION AND CERTIFICATION

During construction, the installation contractor should provide ongoing equipment specifications and pressure testing results along with regular installation updates including evidence of the installer's competency and compliance with the design engineers specification and drawings. These are critical documents that provide confirmation that the work is in accordance with the specification and is being installed to meet the design engineer's documentation.

A regular site inspection by the design engineers, of the installed work, that the equipment being held on site is protected and that the staff is providing the quality of work is necessary to provide an installation that will be suitable for the provision of UHP gases.

Cleaning specifications vary around the world and it's the designer's responsibility to provide a suitable specification to meet either an international or nationally recognized standard or provide an adequate specification that is acceptable for the project. If this is not possible the alternative is to require independent certification that the gases in the pipelines are of the same level of purity that was supplied by the gas manufacturer and that has been connected to the pipeline prior to handover to the laboratory. It is the design engineer's responsibility to ensure the specification and drawings supplied are suitable for the requirements of the facility carrying out the works.

Laboratory gas manufacturer's purity standards for laboratory grade gases are easily available and can be used as guidelines of what purity is provided in their gas cylinders. It is possible to have any gas pipeline tested at the terminal units, a sample is taken for testing for purity by independent recognized laboratories when the systems have been fully tested and commissioned. This would be carried out using the client's gas cylinders connected to the system and online and has been certified as ready for use.

The final commissioning and handover certification must provide suffi-cient information to the client to confirm that the system is in accordance with the design specification and drawings. The certification should also include test results that have been witnessed by an authorized representative of the laboratory, the design engineer and a project manager representing the installation contractor, confirming that the following items have been carried out and passed for use. Each of these items should be tested and recorded as a future reference for the operation of the system:

- Design calculations using a recognized formula to prove plant and equipment are capable of meeting the requirements of the labora-tory for pressure, flow, and cleanliness.
- The gas pipeline design calculations are supplied indicating the flows and pressure losses that have been calculated for the design flow rates specified. This can only be done theoretically and the results of all calculations should be provided.
- All regulatory, engineering, equipment, and laboratory references used to design the system are supplied as part of the handover documentation.
- The materials of construction are compatible with the specific gases installed.
- The equipment installed is in accordance with the specification and is approved by the design engineer and the manufacturers as suitable for the design pressures, flows, and is compatible with the specific gases.
- That each gas supply manifold is fully functional and suitable for the purity and flows specified.
- The gas alarm systems operate as documented in all locations.
- Any gas sensing equipment systems installed are in accordance with local regulations and have been tested as per the specification and the manufacturer's instructions and all functions including interface controls to fire trip systems and remote and local emer-gency shutdown controls operate correctly.
- The isolation valves supplied in the pipeline isolate the correct respective gases only.
- The controls, wiring system, and all functions operate as specified.
- Each terminal outlet is correctly labeled and connected to the correct gas supply.

- Each gas supply is capable of providing the correct pressure and flow at each terminal outlet.
- All signage to allow safe working conditions for all personnel who may be in the vicinity of any gas supplies, pipelines, and ancillary equipment is in accordance with National or Regional Authorities that are applicable to the storage and use of flammable, toxic, and inert gas plant and equipment.

KEYWORDS

- **basic design principles**
- **client expectations**
- **completion and certification**
- **cryogenic gas pipelines**
- **equipment selection**
- **gas cylinder supply**
- **gas system design**
- **laboratory function**
- **mechanical design considerations**
- **vacuum specific requirements**

CHAPTER 5

GAS DATA: INERT GASES

5.1 INTRODUCTION

The availability of gas specific information that is relevant to the installation of piped laboratory gases is spread throughout the scientific, manufacturing, and engineering community with a little cross-reference between the various areas of expertise.

Laboratories are extremely precise in their function and operation and each has its own requirements that dictate the gases to be used; their purity and the performance required of each gas system. To supplement this there is a large volume of information available from the gas manufacturers, as well as from the equipment manufacturers each of which may vary to meet the individual company's product range. As there is no central source where this information is available to obtain at least a general basis for a laboratory gas system design, we hope to provide a starting point for gas data and equipment in this chapter.

This chapter is a collection of gases specific information. It has been gathered from laboratory gas manufacturers, equipment suppliers, and from the manufacturers of the scientific equipment the laboratories use. Some of it comes from resolving localized issues where plant or equipment failed for obscure reasons or failed due to misinformation that had been copied from previous project designs. Each project requires up to date research into the uses of the gases and the associated equipment.

Common cleanliness issues regularly revolve around the purity of the gases at the point of use or the provision of a contaminated gas purity being supplied after passing through the pipeline. Contamination of the gas stream by the pipeline and equipment is an all too common problem, especially where inexperienced installation companies are employed for the work.

Some of the information in the book is repetitive; this has been done purposefully. Rather than provide generic comments that are relative to a number of gases, we have attempted to provide sufficient information about a specific gas that can be obtained from a single subchapter.

There are a number of values that are represented as approximations. Many of these are related to the contents of cylinders and the filling pressures provided by gas suppliers. These may vary between manufacturers and countries, and as such exact measurements are not possible. All pressure ratings are provided in gauge pressure unless noted otherwise, pressure readings at absolute pressure are followed by A.

This book is not intended to afford a 100% foolproof answer to every question; however, it proposes to provide guidelines for design engineers to locate answers to questions they may have and offer suggestions where to look for them.

5.2   INSTRUMENT AIR/ZERO GRADE

5.2.1   DESCRIPTION

There are a variety of types of instrument air; each has different purities and would have been selected by the facility to meet the demands of the processes and procedures being undertaken. The gas may come under different names, some of which represent the same gas; for example, Synthetic Air and Zero Air may represent the same gas although this is not always the case. It is necessary to determine the type of gas prior to designing the system; each gas type will be made up of different impurities that will determine the type of supply source to be provided. The types of instrument air are basically divided into three sources:

5.2.1.1   COMPRESSOR INSTRUMENT AIR SUPPLY

Instrument air produced from a compressor plant can be defined as atmospheric air that has been compressed to an approximate pressure between 800 and 1,200 kPa from a local on-site compressor plant installation. The instrument air is dehydrated using a desiccant drying system to a pressure dew point of −60°C. The air is filtered to remove particulate matter and through absorbers to remove odors and any oil vapors that may be in the inlet air supply. The air complies with International Standards Organization Standard ISO 8573 Compressed Air, Part 1 Contaminants and Purity Classes, Class 1.1.1.

5.2.1.2   ZERO GRADE OR SYNTHETIC AIR FROM HIGH-PRESSURE GAS CYLINDERS

Instrument air from HP cylinder supply source is supplied at a pressure between 15,000 to 30,000 kPa (these pressures vary between countries and gas manufacturers) and may have a <5 ppm moisture. This gas is manufactured by mixing nitrogen, and oxygen and the manufacturing process is strictly controlled.

5.2.1.3   ZERO GRADE AIR FROM COMPRESSOR PLANT SUPPLY

It is also possible to obtain Zero Grade instrument air that has been manufactured from a compressor plant. It may contain many of the gaseous impurities such as argon, and the dew point may not be as low as that provided from HP cylinder supplies. The air may also contain atmospheric contaminants that may not be removed using the compressor filtration system, and additional filtration using molecular sieves may be necessary.

5.2.2   HAZARDS

Instrument Air/Zero Grade is a compressed gas that can cause physical injury to the user or other nearby individuals. An uncontrolled pressurized gas flow provides a source of energy that can accelerate particles and other small items to velocities that may cause physical injury to the user or other nearby individuals.

5.2.3   LABORATORY USE

- Instrument Air/Zero Grade is used to calibrate environmental emission monitoring, industrial hygiene monitors, and trace impurity analyzers and as a balance gas for some calibration mixtures.

- Instrument Air/Zero Grade is used as combustion gas for gas chromatography, flame ionization detectors (FID), flame photometric detectors (FPD) and as an alternative to nitrous oxide in atomic absorption spectrometers (AAS) [23] and for pneumatic controls in these instruments.

5.2.4 REFERENCES

Material Safety Data Sheets are available from the gas manufacturers that provide detailed information on this gas. The information required for any laboratory gas design must include all of the relevant properties of the gas, any compatibility issues that are applicable to the design and the acceptable pressure, flow, concentration, and other pertinent factors that are applicable to the design [19, 21, 22, 24, 28, 31, 33, 58, 60–62].

Many lubricants are compatible with Instrument Air/Zero Grade; however, it is not recommended that these be used in parts of a pipeline that are in direct contact with the gas. If it is necessary to use lubricants, the manufacturers must be contacted to confirm their products are suitable for the intended purpose and will not contaminate the purity of the Instrument Air/Zero Grade air. Care must be taken to ensure that any volatile lubricants or other contaminants that may enter the gas stream and contaminate the gas supply purity are specifically excluded.

Note: The use of lubricants is not recommended for use in any laboratory gas pipelines.

5.2.5 MATERIALS OF CONSTRUCTION

5.2.5.1 COPPER TUBE

Copper tube for use with Instrument Air should be as supplied for medical grade gases in an oxygen clean state that is suitable for the majority of instrument air pipelines [19, 61]. The tube must be delivered to the site pre-cleaned, dehydrated, and capped ready for use. During construction, the tube must be kept off the floor and in a location that prevents contamination from particulates or fluids that may occur on the job site. Medical Gas Grade copper tube is used in the majority of laboratory gas pipelines [19, 61].

It is recommended that hard drawn or semi-hard drawn copper tube is used for laboratory gas pipelines unless there are incompatibility issues or it is recommended otherwise. The use of soft annealed copper pipe is not recommended for laboratory gas pipelines.

5.2.5.2 STAINLESS STEEL TUBE

Used for special instances; however, there is no advantage in the use of stainless steel unless there is a specific compatibility or cleanliness issue that may corrupt the gas supply. The pipeline design must not impact the gas in any way and thereby ensure the gas purity levels are unchanged when delivered at the outlet connection point.

Stainless steel is recommended for high-pressure service where the copper tube is not suitable due to pressure limitations or availability.

5.2.5.3 ELECTROPOLISHED STAINLESS STEEL TUBE

The electropolished stainless steel tube is produced specifically for the electronics industry. It should be noted that this tube should have the same cleanliness level as certified oxygen clean tube.

The electropolished stainless steel tube is used to assist in the prevention of contamination due to outgassing where spontaneously combustible gases are piped and gases may be trapped in the pipe surface.

Note: It has been suggested that the use of this tube type will improve the flow characteristics of a pipeline. However, calculations using the Darcy Weisbach formula [21, 22, 24, 58] to calculate critical pipeline flows indicate that any increase that may occur is of little or no consequence in a laboratory gas pipeline. Any increase in flow that may be provided in using this tube is less than 0.1% of the design flow rate, the use of oxygen cleaned unpolished stainless steel tube of the same size will not impact on the working pressure or design flow. The design flow and system pressure should use a design safety margin normally set at 10% above the system working parameters.

5.2.5.4 BRASS

Brass is used in the manufacture of isolation and control valves. It is also the most common type used in the production of pressure regulators in the majority of cases for gas purity levels up to Ultra High Purity at 99.999%.

Some laboratory bench mounted tapware is manufactured from brass with an external protective coating.

All valves used for the control of laboratory gas pipelines must be oxygen cleaned equal to the national standard for medical grade gases as a minimum; certification of cleanliness levels should be provided by the manufacturer.

5.2.5.5 SYNTHETIC TUBE

The synthetic tube has been suggested for use in high and low-pressure applications. If the synthetic tube is used for high-pressure service it must be rated for the maximum working pressure of the cylinders or pipeline and provide safe working conditions for the laboratory staff. It is used in some high-pressure hoses for cylinder connections to manifolds; however, this tubing is reinforced with stainless steel outer braiding or similar protection.

For low-pressure applications synthetic tube has a different set of problems, these include:

- The emission of volatile organic compounds used in the manufacture of the tube.
- The emission of esters used in the manufacture of the tube.
- The material used in some synthetic tubing is hygroscopic.
- Synthetic tubing and fittings may not be available in an oxygen clean state.
- Synthetic tubing may be porous to some gases, helium, and mixtures containing it are subject to this effect.

Note: The use of synthetic tube is not recommended for laboratory gas pipelines, the tubing may have materials concerns that could contaminate the gas stream. If specifically required the suppliers should provide confirmation that the material is fit for purpose.

5.2.6 PRESSURE REQUIREMENTS

For the majority of laboratory instrument air systems, a pipeline pressure of 410 kPa will be sufficient for the operation of pneumatic controls however there are some specific procedures that will require higher pressures. It is highly recommended that consultation with the laboratory staff is undertaken to ensure that the gas being supplied will meet with their requirements and the equipment being used. There are continually new procedures and equipment that may require gases provided in a unique manner that is not able to be met by the standard laboratory gas pipeline systems.

The working pressure of the pipeline will determine what types and models of valves will need to be used as standard laboratory tapware may vary in operating pressures between 700 and 1,000 kPa. High-pressure systems may require specialized pipeline materials, valves, and tapware that may have to be individually specified to meet these requirements; advice from the manufacturers will be necessary for this type of installation and the design criteria will need to be detailed in any specification to provide information on the proposed working environment.

5.2.7   GAS SUPPLY REQUISITES

There are a number of options available for the supply of gases in laboratories. The factors that determine which type of supply source will be optimal will depend on a number of elements.

5.2.7.1   SYSTEM FLOW DEMAND

Every system will need to be designed to suit the equipment being used in the laboratory. For example, a small single laboratory using one bench mounted gas chromatograph may require a hydrogen and nitrogen or helium supply which, due to the very low flow demand, may be supplied from single cylinders that can be monitored on a weekly basis. As the demand for this type of equipment may be in the range of 50 to 200 mL/min, a 7,000-liters cylinder of gas would provide gas for a number of weeks. A major laboratory would require considerably higher flows and it is imperative to have some information on the types and flow demand of the various items of equipment.

For instrument air the flow rate may vary dramatically as the uses range from low flows for pneumatic controls to considerably higher flows for an AAS [23] (if the AAS is using instrument air instead of nitrous oxide) especially if the unit itself is a large model that may be called upon to be in use for periods in excess of 8 hours.

5.2.7.2   SYSTEM PRESSURE

The pressure requirements will vary to suit the equipment being used in the laboratory. Most manufacturers of laboratory equipment will provide a built-in pressure regulator that allows a range of supply pressures to be tolerated. This must be confirmed by the equipment manufacturer.

If there is a specific pressure required it can be provided by a local pressure regulator at the terminal outlet or at a central point in the laboratory if that is convenient. The pipeline, valves, and the pressure regulator must be sized to suit the maximum design flow required.

5.2.7.3   SYSTEM DEPENDABILITY

Each gas system will have requirements that dictate the type of supply required. Many laboratories require a permanent supply that will be online twenty-four hours a day and seven days a week. This can be accomplished relatively easily for most gas supply systems using manifold cylinders or a central instrument air supply provided by compressors. These systems need a service that will provide notification to laboratory staff that the system gas supply is in its standard operating condition or has a depleted gas supply caused by usage or an equipment failure.

To overcome this problem the plant or cylinder manifold is provided with a backup or reserve supply. In the case of cylinder supply, the number of cylinders required is duplicated with an alarm function to automatically advise staff that the change from main to reserve supply has occurred. For a compressor installation, the equipment is duplicated and any problems with the plant should be remotely monitored.

5.2.8   GAS SUPPLY SYSTEMS

5.2.8.1   AUTOMATIC MANIFOLDS

Automatic manifolds [29] provide a continuous supply of gas that is connected to two independent banks of gas cylinders. Each bank should have the ability to automatically provide a continuous gas supply to the laboratory to allow for a replacement of the in-use cylinder bank while operating from the reserve supply. This will need to take into consideration the time necessary to purchase the replacement cylinders and to have them delivered to the laboratory ready for use.

Purity levels specified by the laboratory must be met by the manifold supplier to ensure compatibility with the gas being piped. Certification should be sought from the recommended supplier that the specified purity level can be provided.

- Automatic manifolds are used in standard installations with maximum pressure requirements that do not exceed 1,000 kPa. For higher pressures, it is recommended that the manufacturer is contacted to confirm what modification is required, is available and is compatible with the pressures specified and that the manifold will meet the expected system demand.
- Automatic changeover manifolds are used where continuity of supply is mandatory.

5.2.8.2   MANUAL MANIFOLDS

Manual manifolds provide a continuous supply of gas that is connected to two independent banks of gas cylinders. Each bank is manually isolated and provides a continuous gas supply to the laboratory to allow for a replacement of the in-use cylinder bank while operating from the reserve supply. This will need to take into consideration the time necessary to purchase the replacement cylinders and to have them delivered to the laboratory ready for use. The changeover from the in use to the reserve cylinders is a manual operation and requires visual inspection of the gas supply on a regular basis.

- Used where monitoring of the gas supply is not critical or can be manually operated and visually monitored; these manifolds require a manual operation to alternate the supply side of the manifold.

5.2.8.3   SINGLE CYLINDER INSTALLATIONS

Small installations that do not require large volumes of gas may also be used for systems that can be manually operated and visually monitored.

- Used when Certified or Gravimetric gases are required with limited use and cost can be a consideration.

5.2.8.4 BULK SUPPLY ALTERNATIVES

Cylinder supply is the common method of providing systems with laboratory gases. They are available in cylinder packs containing up to fifteen cylinders that are interconnected and use a single or dual connection. The manifold and associated equipment used is similar to that used for other laboratory gases however cylinder valve connections may be gas specific; so it is necessary to specify the exact gas and the number of cylinders that are expected to be required.

When using large numbers of cylinders it is necessary to specify the pressure and maximum flow rate for the pressure regulators to meet the system design flow rate.

The supply of instrument air can also be provided by a compressor plant, details of plant configuration and design are discussed in Chapter 2.

5.2.9 OUTLET VALVE TYPES/USES

- Laboratory taps can be either diaphragm (stainless steel diaphragm) or needle type suitably cleaned to the purity level of the gas supply.
- Double olive compression fittings are the most common connection type fitted to the outlets of the taps. Push on or serrated barb connections are unsuitable for the pressures encountered for the laboratory use of this gas.
- Valve manufacturers should be requested to confirm their range of purity levels available for the standard range of sizes.
- Needle type valves with double olive compression fittings are common where high and low-pressure connections are required.
- It is necessary to specify that the materials of construction including all components, seals, and seats are compatible with the gases being piped.
- The inclusion of laboratory taps of unknown cleanliness or purity and material compatibility should be avoided at all costs.
- Medical gas outlet connections are not suitable for laboratory service.

5.2.10 PURITY LEVEL

- Instrument Grade/Zero Air. Refer to equipment manufacturers documentation for proposed purity levels to ensure compliance with the laboratories specified requirements and obtain certified compliance.

5.2.11 ADDITIONAL DESIGN ASSISTANCE REFERENCES

- Adiabatic compression should be avoided wherever possible; the use of slow opening valves is recommended. Gay Lussac's law states that the pressure of a given amount of gas held at constant volume is directly proportional to the temperature in degrees Kelvin. When calculating the physical characteristics of any gas this may be of importance particularly where oxidizing, high pressure, toxic, non-respirable, and flammable gases are being installed.
- If solenoid valves are necessary, the use of slow opening valve types is recommended. These can be controlled from local manually operated emergency push buttons and also interconnected to the fire trip system or any localized emergency gas monitoring system installed in the laboratory.
- Controlled pre-pressurized systems are an optional method of preventing adiabatic compression that can be caused by sudden activation of quick acting valves such as solenoid and ball valves.
- ASTM produces a number of documents that provide technical assistance for the design of gas systems. The ASTM G series publications [14–18] are of particular assistance for these types of installations.
- International and National codes of practice and regulations governing the use of dangerous goods must be referenced. In some countries, these regulations are protected by law and may not be ignored.

5.3 ARGON

5.3.1 DESCRIPTION

Gaseous argon represents 0.94% of the earth's atmosphere. The gas is distilled from the liquefied air during the manufacture of cryogenic liquids. It can be supplied in cylinders at a pressure between 15,824 and 30,000 kPa (these pressures vary between countries and gas manufacturers) or as a cryogenic liquid at pressures up to 1,000 kPa. It is a colorless, tasteless, and odorless inert gas, it is non-flammable.

5.3.2 HAZARDS

- Argon is an asphyxiant that can cause nausea and vomiting followed by a loss of consciousness and death. Loss of consciousness may occur without warning in areas of oxygen concentration below 19%. Personnel may not be physically aware of the effects of the gas and may become unconscious without warning. It is highly recommended that locations using this gas have permanent gas monitoring systems with audible and visual alarms fitted, the provision of warning signs in the area where non-respirable gases are installed is recommended.
- Argon is a compressed gas that can cause physical injury to the user or other nearby individuals. An uncontrolled pressurized gas flow provides a source of energy that can accelerate particles and other small items to velocities that may cause physical injury to the user or other nearby individuals.

5.3.3 LABORATORY USE

- Used in UHP grade mixtures for analyses and quality control.
- UHP argon is used as a plasma gas in Inductive Coupled Plasma (ICP) emission spectrometry [23], blanket gas in graphite furnace atomic absorption spectrometry (GFAAS) [23], and as a carrier gas in gas chromatography (GC and LCGC) [23] for various detectors.
- It is used in a mixture of 10% methane in argon (P10) and is used in Geiger counter calibration and in the detector of X-Ray Fluorescence (XRF) as a quenching gas.

5.3.4 REFERENCES

Material Safety Data Sheets are available from the gas manufacturers that provide detailed information on this gas. The information required for any laboratory gas design must include all of the relevant properties of the gas, any compatibility issues that are applicable to the design and the acceptable pressure, flow, concentration, and other pertinent factors that are applicable to the design [19, 21, 22, 24, 28, 31, 33, 58, 60–62].

Many lubricants are compatible with argon; however, it is not recommended that these be used in parts of a pipeline that is in direct contact with the gas stream. If necessary to use lubricants the manufacturers must be contacted to confirm their products are suitable for the intended purpose and will not contaminate the purity of the argon. Care must be taken to ensure that any volatile lubricants or other contaminants that may enter the gas stream and contaminate the gas supply purity are specifically excluded.

5.3.5   MATERIALS OF CONSTRUCTION

5.3.5.1   COPPER TUBE

Copper tube for use with Argon should be as supplied for medical grade gases in an oxygen clean state is suitable for the majority of argon gas pipelines [19, 61]. The tube must be delivered to site pre-cleaned, dehydrated, and capped ready for use. During construction, the tube must be kept off the floor and in a location that prevents contamination from particulates or fluids that may occur on the job site. Medical Gas Grade Copper tube is used in the majority of laboratory gas pipelines [19, 61].

It is recommended that hard drawn or semi-hard drawn copper tube is used for laboratory gas pipelines unless there are incompatibility issues or it is recommended otherwise by qualified engineers. The use of soft annealed copper pipe is not recommended for laboratory gas pipelines.

5.3.5.2   STAINLESS STEEL

Used for special instances; however, there is no advantage in the use of Stainless steel unless there is a specific compatibility or cleanliness issue that may corrupt the gas supply. The pipeline design must not impact the gas in any way and thereby ensure the gas purity levels are unchanged when delivered at the outlet connection point.

Stainless steel is recommended for high-pressure service where the copper tube is not suitable due to pressure limitations or availability.

5.3.5.3   ELECTROPOLISHED STAINLESS STEEL TUBE

The electropolished stainless steel tube is produced specifically for the electronics industry. It should be noted that this tube should have the same cleanliness level as certified oxygen clean tube.

The electropolished stainless steel tube is used to assist in the prevention of contamination due to outgassing where spontaneously combustible gases are piped and gases may be trapped in the pipe surface.

Note: It has been suggested that the use of this tube type will improve the flow characteristics of a pipeline. However, calculations using the Darcy Weisbach formula [21, 22, 24, 58] to calculate critical pipeline flows indicate that any increase that may occur is of little or no consequence in a laboratory gas pipeline. Any increase in flow that may be provided in using this tube is less than 0.1% of the design flow rate, the use of oxygen cleaned unpolished stainless steel tube of the same size will not impact on the working pressure or design flow. The design flow and system pressure should use a design safety margin normally set at 10% above the system working parameters.

5.3.5.4   BRASS

Brass is used in the manufacture of isolation, and control valves. It is also the most common type used in the production of pressure regulators in the majority of cases for gas purity levels up to ultra-high purity at 99.999%. Some laboratory bench mounted tapware is manufactured from brass with an external protective coating.

All valves used for the control of laboratory gas pipelines must be oxygen cleaned equally to the national standard for medical grade gases as a minimum; certification of cleanliness levels should be provided by the manufacturer.

5.3.5.5   SYNTHETIC TUBE

The synthetic tube has been suggested for use in high and low-pressure applications. If the synthetic tube is used for high-pressure service it must be rated for the maximum working pressure of the cylinders or pipeline and provide safe working conditions for the laboratory staff. It is used in some high-pressure hoses for cylinder connections to manifolds; however, this tubing is reinforced with stainless steel outer braiding or similar protection.

For low-pressure applications synthetic tube has a different set of problems, these include:

- The emission of volatile organic compounds used in the manufacture of the tube.
- The emission of esters used in the manufacture of the tube.
- The material used in some synthetic tubing is hygroscopic.
- Synthetic tubing and fittings may not be available in an oxygen clean state.
- Synthetic tubing may be porous to some gases, helium, and mixtures containing it are subject to this effect.

Note: The use of synthetic tube is not recommended for laboratory gas pipelines; the tubing may have materials that could contaminate the gas stream. If specifically required the suppliers should provide confirmation that the material is fit for purpose.

5.3.6   PRESSURE REQUIREMENTS

For the majority of laboratory argon systems, a pipeline pressure of 410 kPa will be sufficient. It is highly recommended that consultation with the laboratory staff is undertaken to ensure that the gas being supplied will meet with their requirements and the equipment being used. There are continually new procedures and equipment that may require gases provided in a unique manner that is not able to be met by the standard laboratory gas pipeline systems.

The working pressure of the pipeline will determine what types and models of valves will need to be used as standard laboratory tapware may vary in operating pressures between 700 and 1,000 kPa.

5.3.7   GAS SUPPLY REQUISITES

There are a number of options available for the supply of gases in laboratories. The factors that determine which type of supply source will be optimal will depend on a number of elements.

5.3.7.1   SYSTEM FLOW DEMAND

Every system will need to be designed to suit the equipment being used in the laboratory, e.g., a small single laboratory using one ICP [23] may require an argon supply which, due to the flow demand, may be supplied from single cylinders that can be monitored on a daily basis. However, it is usual practice to provide an automatic manifold arrangement with the number of cylinders attached relative to the expected system demand. A major laboratory would require higher flows and it is imperative to have some information on the types of the various items of equipment and their specific flow demand. For argon, pressure, and flow rate may vary dramatically between items of equipment.

5.3.7.2   SYSTEM PRESSURE

The pressure requirements will vary to suit the equipment being used in the laboratory, most manufacturers of laboratory equipment will provide a built-in pressure regulator that allows a range of supply pressures to be tolerated. This must be confirmed by the equipment manufacturer.

   If there is a specific pressure required it can be provided by a local pressure regulator at the terminal outlet or at a central point in the laboratory if that is convenient. The pipeline, valves, and the pressure regulator must be sized to suit the maximum design flow required.

5.3.7.3   SYSTEM DEPENDABILITY

Each gas system will have requirements that dictate the type of supply required; many laboratories require a permanent supply that will be online twenty-four hours a day and seven days a week. This can be accomplished relatively easily for most gas supply systems using manifold cylinders. These systems need a service that will provide notification to laboratory staff that the system gas supply is in its standard operating condition or has a depleted gas supply caused by usage or an equipment failure.

   To overcome this problem the plant or cylinder manifold is provided with a backup or reserve supply, in the case of cylinder supply, the number of cylinders required is duplicated with an alarm function to automatically advise staff that the change from the online supply to the reserve has occurred.

5.3.8   GAS SUPPLY SYSTEMS

5.3.8.1   AUTOMATIC MANIFOLDS

Automatic manifolds [29] provide a continuous supply of gas that is connected to two independent banks of gas cylinders. Each bank should have the ability to automatically provide a continuous gas supply to the laboratory to allow for a replacement of the in-use cylinder bank while operating from the reserve supply. This will need to take into consideration the time necessary to purchase the replacement cylinders and to have them delivered to the laboratory ready for use.

Purity levels specified by the laboratory must be met by the manifold supplier to ensure compatibility with the gas being piped. Certification should be sought from the recommended supplier that the specified purity level can be provided.

- Automatic manifolds are used in standard installations with maximum pressure requirements that do not exceed 1,000 kPa. For higher pressures, it is recommended that the manufacturer is contacted to confirm what modification is required, is available and is compatible with the pressures specified and that the manifold will meet the expected system demand.
- Automatic changeover manifolds are used where continuity of supply is mandatory.

5.3.8.2   MANUAL MANIFOLDS

Manual manifolds provide a continuous supply of gas that is connected to two independent banks of gas cylinders. Each bank is manually isolated and provides a continuous gas supply to the laboratory to allow for a replacement of the in-use cylinder bank while operating from the reserve supply. This will need to take into consideration the time necessary to purchase the replacement cylinders and to have them delivered to the laboratory ready for use. The changeover from the in use to the reserve cylinders is a manual operation and requires visual inspection of the gas supply on a regular basis.

- Used where monitoring of the gas supply is not critical or can be manually operated and visually monitored. These manifolds require a manual operation to alternate the supply side of the manifold.

5.3.8.3   SINGLE CYLINDER INSTALLATIONS

Small installations that do not require large volumes of gas may also be used for systems that can be manually operated and visually monitored.

- Used when certified or gravimetric gases are required and cost can be a consideration.

5.3.8.4   BULK SUPPLY SYSTEMS

5.3.8.4.1   Cylinder Packs

Cylinder supply is the common method of providing systems with laboratory gases. They are available in cylinder packs containing up to fifteen cylinders that are interconnected and use a single or dual connection. The manifold and associated equipment used is similar to that used for other laboratory gases; however, cylinder valve connections may be gas specific; so it is necessary to specify the cylinder valve CGA number [20] and the number of cylinders that are required.

When using large numbers of cylinders, it is necessary to specify the pressure and maximum flow rate for the pressure regulators to meet the system design flow rate.

5.3.8.4.2   Cryogenic Supply

Liquid argon is a cryogenic liquid gas and is available for high demand systems, cryogenic vessels are available in capacities ranging from 3,000 to 25,000 liters from the majority of the gas suppliers. In most instances, the vessels are leased to laboratories under a contract between the gas manufacturer and the laboratory. A cryogenic vessel is a self-contained unit that includes all controls and pressure monitoring equipment as well as a heat exchanger for conversion of the liquid to its gas phase. Liquid argon supply is used where high flows over extended times or high flow applications are required or distances between the manufacturer's source of supply to the location of the laboratory or facility are significant.

5.3.8.4.3 Additional Services Required for Cryogenic Vessel Installations

The use of bulk vessels requires additional services and supplies including:

- 240 V power supply. Or as per national or local services.
- 415 V 32-amp power supply. Or as per national or local services.
- Lighting.
- Security fencing.
- Water connection.
- Road access for a semi-trailer for delivery of liquid supplies.
- Reinforced concrete slab to be located as near to the road as possible to enable the liquid argon delivery tanker easy access for filling purposes. The reinforced concrete slab for the support of the vessel will need to be designed by a structural engineer due to the weight of the vessel when full. The gas manufacturer may be able to assist with this.
- Alarm wiring or telemetry connections from the vessel control to a local monitoring panel may be required to monitor the vessel contents. This may also be connected to the gas supplier's depot for automatic ordering if it is available.
- Compliance with the National or Regional Authorities with suitable licenses obtained where applicable.
- Pressure regulation control panels may be supplied by the gas manufacturer at the source; however, that may not be the case in all locations and this equipment may need to be provided by the installation subcontractor. If this is the case a fully detailed specification in the design document will be necessary.

5.3.9   OUTLET VALVE TYPES/USES

- Laboratory taps can be either diaphragm (stainless steel diaphragm) or needle type suitably cleaned to the purity level of the gas supply.
- Double olive compression fittings are the most common type fitted to the outlets of the taps. Push on or serrated barb connections are unsuitable for the pressures encountered for the laboratory use of this gas.
- Valve manufacturers should be requested to confirm their range of purity levels available for the standard range of sizes.
- Needle type valves with double olive compression fittings are common where high and low-pressure connections are required.
- It is necessary to specify that the materials of construction including all components, seals, and seats are compatible with the gases being piped.
- The inclusion of laboratory taps of unknown cleanliness or purity and material compatibility should be avoided at all costs.
- Medical gas outlet connections are not suitable for laboratory service.

5.3.10   PURITY LEVEL

- Ultra-high purity: 99.999%.

5.3.11   ADDITIONAL DESIGN ASSISTANCE REFERENCES

- Adiabatic compression should be avoided wherever possible; the use of slow opening valves is recommended. Gay Lussac's law states that the pressure of a given amount of gas held at constant volume is directly proportional to the temperature in degrees Kelvin. When calculating the physical characteristics of any gas this may be of importance particularly where oxidizing, high pressure, toxic, non-respirable, and flammable gases are being installed.
- If solenoid valves are necessary the use of slow opening valve types is recommended. These can be controlled from local manually operated emergency push buttons and also interconnected to the

fire trip system or any localized emergency gas monitoring system installed in the laboratory.

- Controlled pre-pressurized systems are an optional method of preventing adiabatic compression that can be caused by sudden activation of quick acting valves such as solenoid and ball valves.
- ASTM produces a number of documents that provide technical assistance for the design of gas systems. The ASTM G-series publications are of particular assistance for these types of installations.
- International and National codes of practice and regulations governing the use of dangerous goods must be referenced. In some countries, these regulations are protected by law and may not be ignored.

5.4   CARBON DIOXIDE: GASEOUS

5.4.1   DESCRIPTION

Carbon dioxide represents 0.03% of the earth's atmosphere; the gas is produced by respiration and the decay of organic compounds and burning carbon. It can be supplied in cylinders at a pressure of 5,824 kPa or as a refrigerated liquid at pressures of approximately 1,000 kPa; it can be used as a refrigerant; however, it is not a cryogenic gas. It is a colorless, tasteless, and odorless inert gas, it is non-flammable.

Carbon dioxide is stored in the cylinder as a liquid at an ambient temperature which weighs considerably more than the same size cylinder containing nitrogen or argon. A gas cylinder containing carbon dioxide contains approximately 2.5 times the volume of gas that a standard gas cylinder contains, e.g., a cylinder of UHP nitrogen contains approximately 7,000 liters of gas whereas a cylinder of carbon dioxide of the same size contains approximately 17,000 liters of gas.

5.4.2   HAZARDS

- Carbon dioxide is a compressed gas where an uncontrolled pressurized gas flow may provide a source of energy that can accelerate particles and other small items to velocities that may cause physical injury to the user or other nearby individuals.

- Carbon dioxide is an asphyxiant that can cause nausea and vomiting followed by a loss of consciousness and death. Loss of consciousness may occur in areas of oxygen concentration below 19%.
- It is highly recommended that locations using this gas have permanent gas monitoring systems with local and remote audible and visual alarms fitted, the provision of warning signs in the area where flammable, toxic, and non-respirable gases are installed is recommended.
- The use of carbon dioxide specific gas sensors is necessary due to the minimum percentage of carbon dioxide in the atmosphere that is considered harmful and any increase in concentration would not be sufficient to actuate a low-level oxygen alarm.

5.4.3   LABORATORY USE

Supercritical carbon dioxide is the mobile phase in both chromatography and extraction applications. High-pressure carbon dioxide is used as a refrigeration agent to precool chambers in certain gas chromatographs.

5.4.4   REFERENCE

Material Safety Data Sheets are available from the gas manufacturers that provide detailed information on this gas. The information required for any laboratory gas design must include all of the relevant properties of the gas, any compatibility issues that are applicable to the design and the acceptable pressure, flow, concentration, and other pertinent factors that are applicable to the design [19, 21, 22, 24, 28, 31, 33, 58, 60–62].

Many lubricants are compatible with carbon dioxide, however, equipment manufacturers must be contacted to confirm their equipment is suitable for purpose and care must be taken to ensure that any selected volatile lubricants that may vaporize in the gas stream will not contaminate the gas purity.

Some seat and seal materials are compatible at low pressures however at full cylinder pressure this is not so, care must be taken to ensure any materials are suitable for use at cylinder pressure when designing high-pressure systems.

Some synthetic hose materials are not suitable for long-term exposure to high or low-pressure carbon dioxide. It has been shown that carbon dioxide has the ability to extract plasticizers from some synthetic hose types.

5.4.5 MATERIALS OF CONSTRUCTION

5.4.5.1 COPPER TUBE

Copper tube for use with carbon dioxide should be as supplied for medical grade gases in an oxygen clean state is suitable for the majority of carbon dioxide gas pipelines [19, 61]. The tube must be delivered to site pre-cleaned, dehydrated, and capped ready for use. During construction, the tube must be kept off the floor and in a location that prevents contamination from particulates or fluids that may occur on the job site. Medical Gas Grade Copper tube is used in the majority of laboratory gas pipelines [19, 61], all pipe used in laboratory gas service must be cleaned for oxygen service.

The use of copper for the majority of laboratory pipelines where the gas stream comes into direct contact with live cells such as in the manipulation of embryonic stem cells, IVF procedures or similar usages copper is not recommended [2]. When the gas is in direct contact with living cells or the gas is used for atmospheric pH control in enclosed atmospheres such as in incubators the use of stainless steel is recommended.

Copper tube is recommended for use in high-pressure service up to 5,824 kPa. The tube wall thickness, valves, and controls must be suitable for the increased working pressure of the system. It is recommended that hard drawn or semi-hard copper tube is used for laboratory gas pipelines unless there are incompatibility issues or it is recommended otherwise by qualified engineers. The use of soft annealed copper pipe is not recommended for laboratory gas pipelines.

5.4.5.2 STAINLESS STEEL TUBE

The stainless steel tube is used for some specialized pipelines however there is no great advantage in the use of stainless steel for general use unless live cells are in direct contact with the gas; gas purity levels are unchanged.

If stainless steel is used for the pipeline construction it is recommended that all valves and controls are manufactured from the same material.

- Recommended for use where the gas may come into direct contact with live cells such as IVF procedures and cell manipulation.

5.4.5.3   ELECTROPOLISHED STAINLESS STEEL TUBE

The electropolished stainless steel tube is produced specifically for the electronics industry. It should be noted that this tube should have the same cleanliness level as certified oxygen clean tube.

The electropolished stainless steel tube is used to assist in the prevention of contamination due to outgassing where spontaneously combustible gases are piped and gases may be trapped in the pipe surface.

Note: It has been suggested that the use of this tube type will improve the flow characteristics of a pipeline. However, calculations using the Darcy Weisbach formula [21, 22, 24, 58] to calculate critical pipeline flows indicate that any increase that may occur is of little or no consequence in a laboratory gas pipeline. Any increase in flow that may be provided in using this tube is less than 0.1% of the design flow rate, the use of oxygen cleaned unpolished stainless steel tube of the same size will not impact on the working pressure or design flow. The design flow and system pressure should use a design safety margin normally set at 10% above the system working parameters.

5.4.5.4   BRASS

Brass is used in the manufacture of isolation and control valves. It is also the most common type used in the production of pressure regulators in the majority of cases for gas purity levels up to ultra-high purity at 99.999%. Some laboratory bench mounted tapware is manufactured from brass with an external protective coating.

All valves used for the control of laboratory gas pipelines must be oxygen cleaned equally to the national standard for medical grade gases as a minimum; certification of cleanliness levels should be provided by the manufacturer.

5.4.5.5   SYNTHETIC TUBE

The synthetic tube has been suggested for use in high and low-pressure applications. If the synthetic tube is used for high-pressure service it must be rated for the maximum working pressure of the cylinders or pipeline and provide safe working conditions for the laboratory staff. It is used in some high-pressure hoses for cylinder connections to manifolds; however, this tubing is reinforced with stainless steel outer braiding or similar protection.

For low-pressure applications synthetic tube has a different set of problems, these include:

- The emission of volatile organic compounds used in the manufacture of the tube.
- The emission of esters used in the manufacture of the tube.
- The material used in some synthetic tubing is hygroscopic.
- Synthetic tubing and fittings may not be available in an oxygen clean state.
- Synthetic tubing may be porous to some gases, helium, and mixtures containing it are subject to this effect.

Note: The use of synthetic tube is not recommended for laboratory gas pipelines, the tubing may have materials concerns that could contaminate the gas stream. If specifically required the suppliers should provide confirmation that the material is fit for purpose.

5.4.6   PRESSURE REQUIREMENTS

For the majority of laboratory carbon dioxide systems, a pipeline pressure of 410 kPa will be sufficient. It is highly recommended that consultation with the laboratory staff is undertaken to ensure that the gas being supplied will meet with their requirements and the equipment being used. There are continually new procedures and equipment that may require gases provided in a unique manner that is not able to be met by a standard laboratory gas pipeline system.

The working pressure of the pipeline will determine what types and models of valves will need to be used as standard laboratory tapware may vary in operating pressures between 700 and 1,000 kPa.

Note 1: High-pressure liquid carbon dioxide systems are dealt with in section 5.5 Carbon Dioxide, High-Pressure Liquid.

Note 2: Safety issue – the expansion of liquid carbon dioxide into its gaseous state results in a 543 to 1 [27] increase in volume (this conversion rate may vary between various Material Safety Data Sheets depending on calculated temperature). All pipelines containing high-pressure liquid carbon dioxide must have safety relief valves with the relief port vented to atmosphere installed at any location where liquid carbon dioxide may be trapped, e.g., between two valves or controls.

5.4.7 GAS SUPPLY REQUISITES

There are a number of options available for the supply of gases in laboratories. The factors that determine which type of supply source will be optimal will depend on a number of elements.

5.4.7.1 SYSTEM FLOW DEMAND

Every system will need to be designed to suit the equipment being used in the laboratory, e.g., a small single laboratory using one bench mounted incubator may require a carbon dioxide supply that may be supplied from a single cylinder that can be monitored on a daily basis. A major laboratory using large incubators would require considerably higher flows and it is imperative to have information on the types and flow demand of the various items of equipment prior to determining the gas supply capacity.

For gaseous carbon dioxide, the flow rate may vary between laboratories as the uses vary for individual locations from less than one liter per minute to central facilities that require flows for multiple large incubators and must be calculated on the number of incubators being installed. The intermittent demand should be taken into consideration as should the maximum time required to return the chamber to the required concentration.

5.4.7.2 SYSTEM PRESSURE

The pressure requirements will vary to suit the equipment being used in the laboratory, most manufacturers of laboratory equipment will provide a built-in pressure regulator that allows a range of supply pressures to be tolerated. This must be confirmed by the equipment manufacturer.

If there is a specific pressure required it can be provided by a local pressure regulator at the terminal outlet or at a central point in the laboratory if that is convenient. The pipeline, valves, and the pressure regulator must be sized to suit the maximum design flow required.

5.4.7.3   SYSTEM DEPENDABILITY

Each gas system will have requirements that dictate the type of supply required, this is no exception. Many laboratories require a permanent supply that will be online twenty-four hours a day and seven days a week. This can be accomplished for most gas supply systems using manifold cylinders or a central gas supply. These systems need a service that will provide notification to laboratory staff that the system gas supply is in its standard operating condition or has a depleted gas supply caused by usage or an equipment failure.

To overcome this problem the cylinder manifold is provided with a backup or reserve supply. For a cylinder supply, the number of cylinders required for operation is duplicated with an alarm function to automatically advise staff that the change from the online supply to the reserve has occurred.

5.4.8   GAS SUPPLY SYSTEMS

5.4.8.1   AUTOMATIC MANIFOLDS

Automatic manifolds [29] provide a continuous supply of gas that is connected to two independent banks of gas cylinders. Each bank should have the ability to automatically provide a continuous gas supply to the laboratory to allow for a replacement of the in-use cylinder bank while operating from the reserve supply. This will need to take into consideration the time necessary to purchase the replacement cylinders and to have them delivered to the laboratory ready for use.

Purity levels specified by the laboratory must be met by the manifold supplier to ensure compatibility with the gas being piped. Certification should be sought from the recommended supplier that the specified purity level can be provided.

- Automatic manifolds are used in standard installations with maximum pressure requirements that do not exceed 1,000 kPa.

For higher pressures, it is recommended that the manufacturer is contacted to confirm what modification is required, is available and is compatible with the pressures specified and that the manifold will meet the expected system demand.

- Automatic changeover manifolds are used where continuity of supply is mandatory.

5.4.8.2   MANUAL MANIFOLDS

Manual manifolds provide a continuous supply of gas that is connected to two independent banks of gas cylinders. Each bank is manually isolated and provides a continuous gas supply to the laboratory to allow for a replacement of the in-use cylinder bank while operating from the reserve supply. This will need to take into consideration the time necessary to purchase the replacement cylinders and to have them delivered to the laboratory ready for use. The changeover from the in use to the reserve cylinders is a manual operation and requires visual inspection of the gas supply on a regular basis.

- Used where monitoring of the gas supply is not critical or can be manually operated and visually monitored. These manifolds require a manual operation to alternate the supply side of the manifold.

5.4.8.3   SINGLE CYLINDER INSTALLATIONS

Small installations that do not require large volumes of gas may also be used for systems that can be manually operated and visually monitored.

- Used when certified or gravimetric gases are required and cost can be a consideration.

5.4.8.4   BULK SUPPLY ALTERNATIVES

5.4.8.4.1   Cylinder Packs

Cylinder supply is the common method of providing systems with laboratory gases. They are available in cylinder packs containing up to fifteen cylinders that are interconnected and use a single or dual connection. The manifold and associated equipment used is similar to that used for other laboratory gases. However, cylinder valve connections may be gas specific; so it is necessary to specify the cylinder valve CGA number [20] and the number of cylinders that are required.

When using large numbers of cylinders it is necessary to specify the use of pressure regulators that are able to meet the system design flow rates.

5.4.8.4.2 *Refrigerated Carbon Dioxide*

Carbon dioxide is available in vacuum insulated vessels similar to the cryogenic vessels noted earlier. The carbon dioxide is stored at a considerably higher temperature than cryogenic gases and has special design considerations that do not apply to those gases. This gas, when stored in this manner, has a maximum flow limitation that is specific to the storage vessel, exceeding this flow rate will cause the vessels internal pressure to decrease and cause the liquid within the vessel to change state to its solid phase, i.e., dry ice. Consultation with the gas manufacturer should be sought to confirm the proposed carbon dioxide design flow rate can be provided by the vessel.

Refrigerated carbon dioxide is available in a range of small vessels, it is necessary to confirm with the local gas suppliers that this type of supply is available in the size and purity required as not all gas manufacturers provide this option. Should these containers be available they may be suitable for refilling in-situ however this must be investigated prior to the design stage as not all gas suppliers provide carbon dioxide in this format.

Vacuum insulated vessels for use with carbon dioxide are similar to cryogenic vessels however they have dissimilar controls and equipment suitable for use with carbon dioxide. It is necessary to contact the local gas supply companies to determine if this option is available.

5.4.8.4.3 *Additional Services Required for Large Bulk Installations*

The use of bulk vessels requires additional services and supplies including:

- Reinforced concrete slab to be located as near to the road as possible to enable the cryogenic delivery tanker easy access for filling purposes. The reinforced concrete slab for the support of the vessel will need to be designed by a structural engineer due to the weight of the vessel when full. The gas manufacturer may be able to assist with this.

- 240 V power supply. Or as per national or local services.
- 415 V 32-amp power supply. Or as per national or local services.
- Lighting.
- Water connection.
- Road access for a semi-trailer for delivery of liquid supplies.
- Security fencing.
- Alarm wiring or telemetry connections from the vessel control to a local monitoring panel may be required to monitor the vessel contents. This may also be connected to the gas supplier's depot for automatic ordering if it is available.
- Pressure regulation control panels may be supplied by the gas manufacturer at the source. However, that may not be the case internationally and this equipment may need to be provided by the installation subcontractor. If this is the case a fully detailed and specified section in the design document will be necessary.

5.4.9   OUTLET VALVE TYPES/USES

- Laboratory taps can be either diaphragm (stainless steel diaphragm) or needle type suitably cleaned to the purity level of the gas supply.
- Double olive compression fittings are the most common connection type fitted to the outlets of the taps. Push on or serrated barb connections are unsuitable for the pressures encountered for the laboratory use of this gas.
- Valve manufacturers should be requested to confirm their range of purity levels available for the standard range of sizes.
- Needle type valves with double olive compression fittings are common where high and low-pressure connections are required.
- It is necessary to specify that the materials of construction including all components, seals, and seats are compatible with the gases being piped.
- The inclusion of laboratory taps of unknown cleanliness or purity and material compatibility should be avoided at all costs.
- Medical gas outlet connections are not suitable for laboratory service.

5.4.10   PURITY LEVEL

- Ultra-high purity: 99.999%.

5.4.11   ADDITIONAL DESIGN ASSISTANCE REFERENCES

- Adiabatic compression should be avoided wherever possible; the use of slow opening valves is recommended. Gay Lussac's law states that the pressure of a given amount of gas held at constant volume is directly proportional to the temperature in degrees Kelvin. When calculating the physical characteristics of any gas this may be of importance particularly where oxidizing, high pressure, toxic, non-respirable, and flammable gases are being installed.
- If solenoid valves are necessary the use of slow opening valve types is recommended. These can be controlled from local manually operated emergency push buttons and also interconnected to the fire trip system or any localized emergency gas monitoring system installed in the laboratory.
- Controlled pre-pressurized systems are an optional method of preventing adiabatic compression that can be caused by sudden activation of quick acting valves such as solenoid and ball valves.
- ASTM produces a number of documents that provide technical assistance for the design of gas systems. The ASTM G series publications [14–18] are of particular assistance for these types of installations.
- International and National codes of practice and regulations governing the use of dangerous goods must be referenced. In some countries, these regulations are enacted in law and may not be ignored.

5.5   CARBON DIOXIDE: HIGH-PRESSURE LIQUID

5.5.1   DESCRIPTION

Carbon dioxide represents 0.03% of the earth's atmosphere; the gas is produced by respiration, and the decay of organic compounds and burning carbon. It can be supplied in cylinders at a pressure of 5,824 kPa or as a low-pressure refrigerated liquid at pressures up to 1,000 kPa; it is not a cryogenic gas. It is a colorless, tasteless, and odorless inert gas, it is non-flammable.

Carbon dioxide is stored in the cylinder as a liquid and weighs considerably more than does a cylinder of the same size containing, for example, gaseous nitrogen. A gas cylinder containing carbon dioxide contains approximately 2.5 times the volume of gas that a standard gas cylinder contains. These cylinders are supplied with a liquid withdrawal facility consisting of a tube between the cylinder valve and the lower part of the gas cylinder. Where high-pressure liquid carbon dioxide is required it is taken through this tube and connected directly to the pipeline for use at full cylinder pressure. It is not possible to reduce the pressure of the high-pressure liquid carbon dioxide. Any pressure regulation will immediately cause the liquid to change state to its gaseous phase.

5.5.2   HAZARDS

- Carbon dioxide at high pressure is a compressed gas that can cause physical injury to the user or other nearby individuals.
- This liquid is piped at full cylinder pressure of 5,824 kPa and must be treated with extreme caution. An uncontrolled pressurized liquid flow provides a source of energy that can accelerate particles and other small items to velocities that may cause physical injury to the user or other nearby individuals.
- Carbon dioxide is an asphyxiant that can cause nausea and vomiting followed by a loss of consciousness and death. Loss of consciousness may occur in areas of oxygen concentration below 19%.
- It is highly recommended that locations using this gas have permanent gas monitoring systems with audible and visual alarms fitted, the provision of warning signs in the area where non-respirable gases are installed is recommended.
- The use of carbon dioxide specific gas sensors is necessary due to the minimum percentage of carbon dioxide in the atmosphere that is considered harmful and any increase in concentration would not be sufficient to actuate a low-level oxygen alarm.

5.5.3   LABORATORY USE

High-pressure liquid carbon dioxide is used as a refrigeration agent to precool sample chambers in certain gas chromatographs.

5.5.4 REFERENCES

Material Safety Data Sheets are available from the gas manufacturers that provide detailed information on this gas. The information required for any laboratory gas design must include all of the relevant properties of the gas, any compatibility issues that are applicable to the design and the acceptable pressure, flow, concentration, and other pertinent factors that are applicable to the design [19, 21, 22, 24, 28, 31, 33, 58, 60–62].

Many lubricants are compatible with carbon dioxide; however, it is not recommended that these be used in parts of a pipeline that is in direct contact with the gas stream. If necessary to use lubricants the manufacturers must be contacted to confirm their products are suitable for the intended purpose and will not contaminate the purity of the carbon dioxide. Care must be taken to ensure that any volatile lubricants or other contaminants that may enter the gas stream and contaminate the gas supply purity are specifically excluded.

In this instance temperature considerations are necessary as the temperature of the cylinders while in storage must be kept below 30°C, temperatures in excess of this will change the state of the high-pressure phase liquid in the cylinder to its gaseous state which will render the system inoperable for laboratory use.

Some seat and seal materials are incompatible with this gas at full cylinder pressure, care must be taken to ensure any seal and seat materials used throughout the installation must be suitable for use at 5,824 kPa.

Some synthetic hose materials are not suitable for long-term exposure to carbon dioxide. It has been shown that this gas has the ability to extract plasticizers from some synthetic hose types.

5.5.5 MATERIALS OF CONSTRUCTION

5.5.5.1 COPPER TUBE

High-pressure liquid carbon dioxide is piped at approximately 5,824 kPa and the proposed maximum working pressure of the copper tube must take this into consideration and suitable wall thicknesses specified.

Note: Copper tube supplied for low-pressure laboratory grade gases in an oxygen clean state is suitable for the majority of laboratory carbon dioxide gas pipelines [19, 61]. However, this may not meet the pressure requirements necessary for high-pressure liquid carbon dioxide.

The tube must be delivered to site pre-cleaned, dehydrated, and capped ready for use. During construction, the tube must be kept off the floor and in a location that prevents contamination from particulates or fluids that may occur on the job site.

It is recommended that hard drawn to be used for laboratory gas pipelines unless there are incompatibility issues or it is recommended otherwise.

5.5.5.2   STAINLESS STEEL

Used for special instances; however, there is no advantage in the use of stainless steel unless there is a specific pressure, compatibility or cleanliness issue that may corrupt the gas supply or have safety issues. The pipeline design must not impact the gas in any way and thereby ensure the gas purity levels are unchanged when delivered at the outlet connection point.

Stainless steel is recommended for high-pressure service where the copper tube is not suitable due to pressure limitations or availability of acceptability of high-pressure copper tube.

5.5.5.3   ELECTROPOLISHED STAINLESS STEEL

The electropolished stainless steel tube is produced specifically for the electronics industry. It should be noted that this tube should have the same cleanliness level as certified oxygen clean tube.

The electropolished stainless steel tube is used to assist in the prevention of contamination due to outgassing where spontaneously combustible gases are piped and gases may be trapped in the pipe surface.

Note: It has been suggested that the use of this tube type will improve the flow characteristics of a pipeline. However, calculations using the Darcy Weisbach formula [21, 22, 24, 58] to calculate critical pipeline flows indicate that any increase that may occur is of little or no consequence in a laboratory gas pipeline. Any increase in flow that may be provided using this tube is less than 0.1% of the design flow rate, the use of oxygen cleaned unpolished stainless steel tube of the same size will not impact on the working pressure or design flow. The design flow and system pressure should use a design safety margin normally set at 10% above the system working parameters.

5.5.5.4   BRASS

Brass is used in the manufacture of isolation and control valves. It is also the most common type used in the production of pressure regulators in the majority of cases for gas purity levels up to ultra-high purity at 99.999%. Some laboratory bench mounted tapware is manufactured from brass with an external protective coating.

All valves used for the control of laboratory gas pipelines must be oxygen cleaned equally to the national standard for medical grade gases as a minimum; certification of cleanliness levels should be provided by the manufacturer.

5.5.5.5   SYNTHETIC TUBE

The synthetic tube has been suggested for use in high and low-pressure applications. If the synthetic tube is used for high-pressure service it must be rated for the maximum working pressure of the cylinders or pipeline. In this case, 5,824 kPa, and provide safe working conditions for the laboratory staff. It is used in some high-pressure hoses for cylinder connections to manifolds; however, this tubing is reinforced with stainless steel outer braiding or similar protection.

Carbon dioxide has been shown to extract the plasticizers from some synthetic tubing, this causes the tube to become brittle and rupture.

The following are concerns that have been found in some synthetic pipeline systems:

- The emission of volatile organic compounds used in the manufacture of the tube.
- The emission of esters used in the manufacture of the tube.
- The material used in some synthetic tubing is hygroscopic.
- Synthetic tubing and fittings may not be available in an oxygen clean state.

Note: The use of synthetic tube is not recommended for laboratory gas pipelines, the tubing may have materials concerns that could contaminate the gas stream. If specifically required the suppliers should provide confirmation that the material is fit for purpose.

5.5.6   PRESSURE REQUIREMENTS

The working pressure of the pipeline is determined by the gas which is taken at full cylinder pressure. Types and models of valves will need to be suitable for operation at the high pressure; standard laboratory tapware with operating pressures between 700 and 1,000 kPa are not suitable for use with this gas.

High-pressure systems will require specialized pipeline materials, valves, and tapware that may have to be individually specified to meet these requirements; advice from the manufacturers will be necessary for this type of installation and the design criteria will need to be detailed in any specification to provide information on the proposed working environment. Valves with dual olive compression fittings are the usual choice, discussions with the suppliers will provide advice on suitable models.

Note: Safety issue – the expansion of liquid carbon dioxide into its gaseous state results in a 543 to 1 [1, 24, 31] increase in volume (this conversion rate may vary between various Material Safety Data Sheets depending on calculated temperature). All pipelines containing high-pressure liquid carbon dioxide must have safety relief valves with the relief port vented to atmosphere installed at any location where liquid carbon dioxide may be trapped, e.g., between two valves or controls.

5.5.7   GAS SUPPLY REQUISITES

For liquid carbon dioxide at 5,824 kPa the options available are limited, the gas cylinders themselves are specifically produced for liquid withdrawal and have a black vertical stripe for the length of the cylinder. The factors that determine which type of supply source will be optimal will depend on a number of elements.

5.5.7.1   SYSTEM FLOW DEMAND

Every system will need to be designed to suit the equipment being used in the laboratory. For high-pressure liquid carbon dioxide, there is a reduced number of options as the usage is extremely low so small bore pipelines are ideally suited for this purpose. A major laboratory would require higher flows; however, due to the intermittent nature of the demand and the use of the specific types of gas chromatographs it would be unusual to require more than a single cylinder supply.

5.5.7.2   SYSTEM PRESSURE

High-pressure liquid carbon dioxide is supplied at a pressure of 5,824 kPa; it cannot be regulated and must be piped to the outlet without pressure adjustment or variation. In its liquid phase high-pressure carbon dioxide remains as a liquid up to a temperature of 31°C at which point it changes to its gaseous phase and the pressure increases accordingly. It cannot be used in the laboratory should this occur and pipelines should be kept clear of direct sunlight or other heat sources.

Under normal operating conditions the pressure is constant until there is no liquid remaining in the cylinder at which time the pressure will reduce rapidly until empty.

5.5.7.3   SYSTEM DEPENDABILITY

This use of liquid carbon dioxide is intermittent and single cylinders that can be physically monitored are recommended unless there is a specific requirement to provide alternative arrangements.

5.5.8   GAS SUPPLY SYSTEMS

5.5.8.1   AUTOMATIC MANIFOLDS

Automatic manifolds are not available for this service.

5.5.8.2   MANUAL MANIFOLDS

Manual manifolds provide a continuous supply of gas that is connected to two independent banks of gas cylinders. Each bank is manually isolated and provides a continuous gas supply to the laboratory to allow for a replacement of the in-use cylinder bank while operating from the reserve supply. This will need to take into consideration the time necessary to purchase the replacement cylinders and to have them delivered to the laboratory ready for use. The changeover from the in use to the reserve cylinders is a manual operation and requires visual inspection of the gas supply on a regular basis.

Manual manifolds are used for multi-cylinder manifolds for liquid carbon dioxide supply; the liquid is drawn from the bottom of the cylinder until empty. Pressure regulators are not used.

5.5.8.3   SINGLE CYLINDER INSTALLATIONS

Small installations that do not require large volumes of gas are suggested for this type of installation.

5.5.8.4   BULK SUPPLY ALTERNATIVES

Not available.

5.5.9   OUTLET VALVE TYPES/USES

- Laboratory taps for high-pressure liquid carbon dioxide should be a needle or regulating stem types suitably cleaned to the purity level of the gas.
- Double olive compression fittings are the most common connection type fitted to the outlets of the taps. Push on or serrated barb connections are unsuitable for the pressures encountered for the laboratory use of this gas.
- Valve manufacturers should be requested to confirm their range of high-pressure capability and purity levels available for the range of sizes required.
- It is necessary to specify that the materials of construction including all components, seals, and seats are compatible with the gas being piped.
- The inclusion of laboratory taps of unknown cleanliness or purity and material compatibility should be avoided at all costs.
- Medical gas outlet connections have maximum working pressures that are not suitable for service with high-pressure liquid carbon dioxide.

5.5.10   PURITY LEVEL

There is only one type of high-pressure liquid carbon dioxide supply; the cylinder is easily identifiable having a vertical black stripe from the top to bottom of the cylinder.

5.5.11   ADDITIONAL DESIGN ASSISTANCE REFERENCES

- High-pressure liquid carbon dioxide has specific design concerns; these include the system pressure and the maximum temperature. The liquid has physical constraints that must be taken into consideration; the information is available from gas manufacturer's websites which indicate the maximum allowable temperature that the pipeline can be exposed to after which the liquid changes state to the gaseous phase with comparable pressure increases.
- Adiabatic compression should not be a problem as the gas is in its liquid phase however suitable protection should be provided to prevent overpressure of the pipeline during use.
- If solenoid valves are necessary the use of slow opening valve types is recommended. These can be controlled from local manually operated emergency push buttons and also interconnected to the fire trip system or any localized emergency gas monitoring system installed in the laboratory.
- Overpressure relief valves should be installed in any pipe section where the liquid carbon dioxide may be trapped.
- ASTM produces a number of documents that provide technical assistance for the design of gas systems. The ASTM G series publications [14–18] are of particular assistance for these types of installations.
- International and National codes of practice and regulations governing the use of dangerous goods must be referenced. In some countries, these regulations are protected by law and may not be ignored.

5.6   HELIUM

5.6.1   DESCRIPTION

Gaseous helium represents 0.0005% of the earth's atmosphere and is recovered from natural gas deposits. It can be supplied in cylinders at a pressure of 14,000 kPa (these pressures vary between countries and gas manufacturers) or as a cryogenic liquid at pressures up to 1,000 kPa. It is a colorless, tasteless, and odorless inert gas, it is non-flammable.

5.6.2 HAZARDS

- Helium is a compressed gas that can cause physical injury to the user or other nearby individuals. An uncontrolled pressurized gas flow provides a source of energy that can accelerate particles and other small items to velocities that may cause physical injury to the user or other nearby individuals.
- Helium is an asphyxiant that can cause nausea and vomiting followed by a loss of consciousness and death. Loss of consciousness may occur without warning in areas of oxygen concentration below 19%. Personnel may not be physically aware of the effects of the gas and may become unconscious without warning.
- It is highly recommended that locations using this gas have permanent gas monitoring systems with audible and visual alarms fitted, the provision of warning signs in the area where non-respirable gases are installed is recommended.

5.6.3 LABORATORY USE

Helium is used as a carrier in gas chromatography. In its liquid state at –269°C, helium is the cooling fluid in magnetic resonance imaging equipment (MRI), magnetic resonance nuclear spectroscopy (NMR), or electron paramagnetic resonance (EPR) magnets.

5.6.4 REFERENCES

Material Safety Data Sheets are available from the gas manufacturers that provide detailed information on this gas. The information required for any laboratory gas design must include all of the relevant properties of the gas, any compatibility issues that are applicable to the design and the acceptable pressure, flow, concentration, and other pertinent factors that are applicable to the design [19, 21, 22, 24, 28, 31, 33, 58, 60–62].

Many lubricants are compatible with helium; however, it is not recommended that these be used in parts of a pipeline that is in direct contact with the gas stream. If necessary to use lubricants the manufacturers must be contacted to confirm their products are suitable for the intended purpose and will not contaminate the purity of the helium. Care must be taken to ensure that any volatile lubricants or other contaminants that may enter the gas stream and contaminate the gas supply purity are specifically excluded.

5.6.5   MATERIALS OF CONSTRUCTION

5.6.5.1   COPPER TUBE

Copper tube for use with helium should be as supplied for medical grade gases in an oxygen clean state is suitable for the majority of helium gas pipelines [19, 61]. The tube must be delivered to site pre-cleaned, dehydrated, and capped ready for use. During construction, the tube must be kept off the floor and in a location that prevents contamination from particulates or fluids that may occur on the job site. Medical Gas Grade Copper tube is used in the majority of laboratory gas pipelines [19, 61].

It is recommended that hard drawn or semi-hard copper tube is used for laboratory gas pipelines unless there are incompatibility issues or it is recommended otherwise. The use of soft annealed copper pipe is not recommended for laboratory gas pipelines.

5.6.5.2   STAINLESS STEEL

Used for special instances; however, there is no advantage in the use of stainless steel unless there is a specific compatibility or cleanliness issue that may corrupt the gas supply. The pipeline design must not impact the gas in any way and thereby ensure the gas purity levels are unchanged when delivered at the outlet connection point. Stainless steel is recommended for high-pressure service where the copper tube is not suitable due to pressure limitations or availability.

5.6.5.3   ELECTROPOLISHED STAINLESS STEEL

The electropolished stainless steel tube is produced specifically for the electronics industry. It should be noted that this tube should have the same cleanliness level as certified oxygen clean tube.

The electropolished stainless steel tube is used to assist in the prevention of contamination due to outgassing where spontaneously combustible gases are piped and gases may be trapped in the pipe surface.

Note: It has been suggested that the use of this tube type will improve the flow characteristics of a pipeline. However, calculations using the Darcy Weisbach formula [21, 22, 24, 58] to calculate critical pipeline flows indicate that any increase that may occur is of little or no consequence in a laboratory gas pipeline. Any increase in flow that may be provided in using this tube is less than 0.1% of the design flow rate, the use of oxygen cleaned unpolished stainless steel tube of the same size will not impact on the working pressure or design flow. The design flow and system pressure should use a design safety margin normally set at 10% above the system working parameters.

5.6.5.4   BRASS

Brass is used in the manufacture of isolation, and control valves. It is also the most common type used in the production of pressure regulators in the majority of cases for gas purity levels up to ultra-high purity at 99.999%. Some laboratory bench mounted tapware is manufactured from brass with an external protective coating.

All valves used for the control of laboratory gas pipelines must be oxygen cleaned equally to the national standard for medical grade gases as a minimum; certification of cleanliness levels should be provided by the manufacturer.

5.6.5.5   SYNTHETIC TUBE

The synthetic tube has been suggested for use in high and low-pressure applications. If the synthetic tube is used for high-pressure service it must be rated for the maximum working pressure of the cylinders or pipeline and provide safe working conditions for the laboratory staff. It is used in some high-pressure hoses for cylinder connections to manifolds; however, this tubing is reinforced with stainless steel outer braiding or similar protection.

For low-pressure applications synthetic tube has a different set of problems, these include:

- Synthetic tubing may be porous to some gases, helium, and mixtures containing it are subject to this effect.
- The emission of volatile organic compounds used in the manufacture of the tube.
- The emission of esters used in the manufacture of the tube.
- The material used in some synthetic tubing is hygroscopic.
- Synthetic tubing and fittings may not be available in an oxygen clean state.

Note: The use of synthetic tube is not recommended for laboratory gas pipelines and in particular for helium. The tubing may have materials concerns that could contaminate the gas stream, if specifically required the suppliers should provide confirmation that the material is fit for purpose.

5.6.6   PRESSURE REQUIREMENTS

For the majority of laboratory helium systems, a pipeline pressure of 410 kPa will be sufficient. It is highly recommended that consultation with the laboratory staff is undertaken to ensure that the gas being supplied will meet with their requirements and the equipment being used. There are continually new procedures and equipment that may require gases provided in a unique manner that is not able to be met by the standard laboratory gas pipeline systems.

The working pressure of the pipeline will determine what types and models of valves will need to be used as standard laboratory tapware may vary in operating pressures between 700 and 1,000 kPa.

5.6.7   GAS SUPPLY REQUISITES

There are a number of options available for the supply of gases in laboratories. The factors that determine which type of supply source will be optimal will depend on a number of elements.

5.6.7.1   SYSTEM FLOW DEMAND

Every system will need to be designed to suit the equipment being used in the laboratory, e.g., a small single laboratory using one bench mounted Gas Chromatograph may require a hydrogen and nitrogen or helium supply which, due to the very low flow demand, may be supplied from single cylinders that can be monitored on a weekly basis. As the demand for this type of equipment may be in the range of 50 to 200 mL/min, a 7,000-liters cylinder of gas would provide gas for a number of weeks. A major laboratory would require higher flows; however, due to the low flow demand at any individual outlet, it is imperative to have some information on the types and flow demand of the various items of equipment being used.

5.6.7.2   SYSTEM PRESSURE

The pressure requirements will vary to suit the equipment being used in the laboratory, most manufacturers of laboratory equipment will provide a built-in pressure regulator that allows a range of supply pressures to be tolerated. This must be confirmed by the equipment manufacturer.

If there is a specific pressure required it can be provided by a local pressure regulator at the terminal outlet or at a central point in the laboratory if that is convenient, the pipeline, valves, and the pressure regulator must be sized to suit the maximum design flow required.

5.6.7.3   SYSTEM DEPENDABILITY

Each gas system will have requirements that dictate the type of supply required; many laboratories require a permanent supply that will be online twenty-four hours a day and seven days a week. This can be accomplished relatively easily for most gas supply systems using manifold cylinders. These systems need a service that will provide notification to laboratory staff that the system gas supply is in its standard operating condition or has a depleted gas supply caused by usage or an equipment failure.

To overcome this problem the plant or cylinder manifold is provided with a backup or reserve supply, in the case of cylinder supply, the number of cylinders required is duplicated with an alarm function to automatically advise staff that the change from the online supply to the reserve has occurred.

5.6.8   GAS SUPPLY SYSTEMS

5.6.8.1   AUTOMATIC MANIFOLDS

Automatic manifolds [29] provide a continuous supply of gas that is connected to two independent banks of gas cylinders. Each bank should have the ability to automatically provide a continuous gas supply to the laboratory to allow for a replacement of the in-use cylinder bank while operating from the reserve supply. This will need to take into consideration the time necessary to purchase the replacement cylinders and to have them delivered to the laboratory ready for use.

Purity levels specified by the laboratory must be met by the manifold supplier to ensure compatibility with the gas being piped. Certification should be sought from the recommended supplier that the specified purity level can be provided.

- Automatic manifolds are used in standard installations with maximum pressure requirements that do not exceed 1,000 kPa. For higher pressures, it is recommended that the manufacturer is contacted to confirm what modification is required, is available and is compatible with the pressures specified and that the manifold will meet the expected system demand.
- Automatic changeover manifolds are used where continuity of supply is mandatory.

5.6.8.2   MANUAL MANIFOLDS

Manual manifolds provide a continuous supply of gas that is connected to two independent banks of gas cylinders. Each bank is manually isolated and provides a continuous gas supply to the laboratory to allow for a replacement of the in-use cylinder bank while operating from the reserve supply. This will need to take into consideration the time necessary to purchase the replacement cylinders and to have them delivered to the laboratory ready for use. The changeover from the in use to the reserve cylinders is a manual operation and requires visual inspection of the gas supply on a regular basis.

- Used where monitoring of the gas supply is not critical or can be manually operated and visually monitored. These manifolds require a manual operation to alternate the supply side of the manifold.

5.6.8.3   SINGLE CYLINDER INSTALLATIONS

Small installations that do not require large volumes of gas may also be used for systems that can be manually operated and visually monitored.

- Used when certified or gravimetric gases are required and cost can be a consideration.

5.6.9   OUTLET VALVE TYPES/USES

- Laboratory taps can be either diaphragm (stainless steel diaphragm) or needle type suitably cleaned to the purity level of the gas supply.
- Double olive compression fittings are the most common type fitted to the outlets of the taps. Push on or serrated barb connections are unsuitable for the pressures encountered for the laboratory use of this gas.
- Valve manufacturers should be requested to confirm their range of purity levels available for the standard range of sizes.
- Needle type valves with double olive compression fittings are common where high and low-pressure connections are required.
- It is necessary to specify that the materials of construction including all components, seals, and seats are compatible with the gases being piped.
- The inclusion of laboratory taps of unknown cleanliness or purity and material compatibility should be avoided at all costs.
- Medical gas outlet connections are not suitable for laboratory service.

5.6.10   PURITY LEVEL

- Ultra-high purity: 99.999%.

5.6.11 ADDITIONAL DESIGN ASSISTANCE REFERENCES

- Adiabatic compression should be avoided wherever possible; the use of slow opening valves is recommended. Gay Lussac's law states that the pressure of a given amount of gas held at constant volume is directly proportional to the temperature in degrees Kelvin. When calculating the physical characteristics of any gas this may be of importance particularly where oxidizing, high pressure, toxic, non-respirable, and flammable gases are being installed.
- If solenoid valves are necessary the use of slow opening valve types is recommended. These can be controlled from local manually operated emergency push buttons and also interconnected to the fire trip system or any localized emergency gas monitoring system installed in the laboratory.
- Controlled pre-pressurized systems are an optional method of preventing adiabatic compression that can be caused by sudden activation of quick acting valves such as solenoid and ball valves.
- ASTM produces a number of documents that provide technical assistance for the design of gas systems. The ASTM G series publications [14–18] are of particular assistance for these types of installations.
- International and National codes of practice and regulations governing the use of dangerous goods must be referenced. In some countries, these regulations are protected by law and may not be ignored.

5.7 NITROGEN: GASEOUS UHP

5.7.1 DESCRIPTION

Gaseous nitrogen represents 78% of the earth's atmosphere; the gas is distilled from the liquefied air during the manufacture of cryogenic liquids. It can be supplied in cylinders at a pressure between 15,824 and 30,000 kPa (these pressures vary between countries and gas manufacturers) or as a cryogenic liquid at pressures up to 1,000 kPa. It is a colorless, tasteless, odorless, inert gas and is an asphyxiant, it is non-flammable.

5.7.2 HAZARDS

- Gaseous nitrogen is an asphyxiant that can cause nausea and vomiting followed by a loss of consciousness and death. Loss of consciousness may occur without warning in areas of oxygen concentration below 19%. Personnel may not be physically aware of the effects of the gas and may become unconscious without warning.
- It is highly recommended that locations using this gas have permanent gas monitoring systems with local and remote audible and visual alarms fitted, the provision of warning signs in the area where flammable, toxic, and non-respirable gases are installed is recommended.
- Gaseous nitrogen is a compressed gas that can cause physical injury to the user or other nearby individuals. An uncontrolled pressurized gas flow provides a source of energy that can accelerate particles and other small items to velocities that may cause physical injury to the user or other nearby individuals.

5.7.3 USES

- Utilized in UHP grade, gaseous nitrogen is used as a carrier gas in gas chromatography for various laboratory and clinical analyses and quality control.
- Gaseous nitrogen is the balance gas of the calibration gas mixtures for environmental monitoring systems and industrial hygiene gas mixtures.
- Gaseous nitrogen is largely used as purge, drier or blanket gas for analyzers or chemical reactors.

5.7.4 REFERENCES

Material Safety Data Sheets are available from the gas manufacturers that provide detailed information on this gas. The information required for any laboratory gas design must include all of the relevant properties of the gas, any compatibility issues that are applicable to the design and the acceptable pressure, flow, concentration, and other pertinent factors that are applicable to the design [19, 21, 22, 24, 28, 31, 33, 58, 60–62].

Many lubricants are compatible with gaseous nitrogen; however, it is not recommended that these be used in parts of a pipeline that is in direct contact with the gas stream. If necessary to use lubricants the manufacturers must be contacted to confirm their products are suitable for the intended purpose and will not contaminate the purity of the gaseous nitrogen. Care must be taken to ensure that any volatile lubricants or other contaminants that may enter the gas stream and contaminate the gas supply purity are specifically excluded.

5.7.5   MATERIALS OF CONSTRUCTION

5.7.5.1   COPPER TUBE

Copper tube for use with nitrogen should be as supplied for medical grade gases in an oxygen clean state is suitable for the majority of nitrogen gas pipelines [19, 61]. The tube must be delivered to site pre-cleaned, dehydrated, and capped ready for use. During construction, the tube must be kept off the floor and in a location that prevents contamination from particulates or fluids that may occur on the job site. Medical Gas Grade copper tube is used in the majority of laboratory gas pipelines [19, 61].

It is recommended that hard drawn or semi-hard drawn copper tube is used for laboratory gas pipelines unless there are incompatibility issues or it is recommended otherwise. The use of soft annealed copper pipe is not recommended for laboratory gas pipelines.

5.7.5.2   STAINLESS STEEL

Used for special instances; however, there is no advantage in the use of stainless steel unless there is a specific compatibility or cleanliness issue that may corrupt the gas supply. The pipeline design must not impact the gas in any way and thereby ensure the gas purity levels are unchanged when delivered at the outlet connection point. Stainless steel is recommended for high-pressure service where the copper tube is not suitable due to pressure limitations or availability.

5.7.5.3   ELECTROPOLISHED STAINLESS STEEL TUBE

The electropolished stainless steel tube is produced specifically for the electronics industry. It should be noted that this tube should have the same cleanliness level as certified oxygen clean tube.

The electropolished stainless steel tube is used to assist in the prevention of contamination due to outgassing where spontaneously combustible gases are piped and gases may be trapped in the pipe surface.

Note: It has been suggested that the use of this tube type will improve the flow characteristics of a pipeline. However, calculations using the Darcy Weisbach formula [21, 22, 24, 58] to calculate critical pipeline flows indicate that any increase that may occur is of little or no consequence in a laboratory gas pipeline. Any increase in flow that may be provided in using this tube is less than 0.1% of the design flow rate, the use of oxygen cleaned unpolished stainless steel tube of the same size will not impact on the working pressure or design flow. The design flow and system pressure should use a design safety margin normally set at 10% above the system working parameters.

5.7.5.4 BRASS

Brass is used in the manufacture of isolation, and control valves. It is also the most common type used in the production of pressure regulators in the majority of cases for gas purity levels up to ultra-high purity at 99.999%. Some laboratory bench mounted tapware is manufactured from brass with an external protective coating.

All valves used for the control of laboratory gas pipelines must be oxygen cleaned equally to the national standard for medical grade gases as a minimum; certification of cleanliness levels should be provided by the manufacturer.

5.7.5.5 SYNTHETIC TUBE

The synthetic tube has been suggested for use in high and low-pressure applications. If the synthetic tube is used for high-pressure service it must be rated for the maximum working pressure of the cylinders or pipeline and provide safe working conditions for the laboratory staff. It is used in some high-pressure hoses for cylinder connections to manifolds; however, this tubing is reinforced with stainless steel outer braiding or similar protection.

For low-pressure applications synthetic tube has a different set of problems, these include:

- The emission of volatile organic compounds used in the manufacture of the tube.
- The emission of esters used in the manufacture of the tube.
- The material used in some synthetic tubing is hygroscopic.
- Synthetic tubing and fittings may not be available in an oxygen clean state.
- Synthetic tubing may be porous to some gases; helium and mixtures containing it are subject to this effect.

Note: The use of synthetic tube is not recommended for laboratory gas pipelines, the tubing may have materials concerns that could contaminate the gas stream, if specifically required the suppliers should provide confirmation that the material is fit for purpose.

5.7.6   PRESSURE REQUIREMENTS

For the majority of laboratory gaseous nitrogen systems, a pipeline pressure of 410 kPa will be sufficient. It is highly recommended that consultation with the laboratory staff is undertaken to ensure that the gas being supplied will meet with their requirements and the equipment being used. There are continually new procedures and equipment that may require gases provided in a unique manner that is not able to be met by the standard laboratory gas pipeline systems.

The working pressure of the pipeline will determine what types and models of valves will need to be used as standard laboratory tapware may vary in operating pressures between 700 and 1,000 kPa. High-pressure systems may require specialized pipeline materials, valves, and tapware that may have to be individually specified to meet these requirements; advice from the manufacturers will be necessary for this type of installation and the design criteria will need to be detailed in any specification to provide information on the proposed working environment.

5.7.7   GAS SUPPLY REQUISITES

There are a number of options available for the supply of gases in laboratories. The factors that determine which type of supply source will be optimal will depend on a number of elements.

5.7.7.1   SYSTEM FLOW DEMAND

Every system will need to be designed to suit the equipment being used in the laboratory, e.g., a small single laboratory using one bench mounted Gas Chromatograph [23] may require a hydrogen and nitrogen or helium supply which, due to the very low flow demand, may be supplied from single cylinders that can be monitored on a weekly basis. As the demand for this type of equipment may be in the range of 50 to 200 mL/min, a 7,000-liters cylinder of gas would provide gas for a number of weeks. A major laboratory would require higher flows; however, due to the low flow demand at any individual outlet, it is imperative to have some information on the types and flow demand of the various items of equipment being used.

5.7.7.2   SYSTEM PRESSURE

The pressure requirements will vary to suit the equipment being used in the laboratory, most manufacturers of laboratory equipment will provide a built-in pressure regulator that allows a range of supply pressures to be tolerated. This must be confirmed by the equipment manufacturer.

 If there is a specific pressure required it can be provided by a local pressure regulator at the terminal outlet or at a central point in the laboratory if that is convenient, the pipeline, valves, and the pressure regulator must be sized to suit the maximum design flow required.

5.7.7.3   SYSTEM DEPENDABILITY

Each gas system will have requirements that dictate the type of supply required; many laboratories require a permanent supply that will be online twenty-four hours a day and seven days a week. This can be accomplished relatively easily for most gas supply systems using manifold cylinders or a central liquid nitrogen supply. These systems need a service that will provide notification to laboratory staff that the system gas supply is in its standard operating condition or has a depleted gas supply caused by usage or an equipment failure.

 To overcome this problem the plant or cylinder manifold is provided with a backup or reserve supply, in the case of cylinder supply, the number of cylinders required is duplicated with an alarm function to automatically advise staff that the change from the online supply to the reserve has occurred.

5.7.8   GAS SUPPLY SYSTEMS

5.7.8.1   AUTOMATIC MANIFOLDS

Automatic manifolds [29] provide a continuous supply of gas that is connected to two independent banks of gas cylinders. Each bank should have the ability to automatically provide a continuous gas supply to the laboratory to allow for a replacement of the in-use cylinder bank while operating from the reserve supply. This will need to take into consideration the time necessary to purchase the replacement cylinders and to have them delivered to the laboratory ready for use.

Purity levels specified by the laboratory must be met by the manifold supplier to ensure compatibility with the gas being piped. Certification should be sought from the recommended supplier that the specified purity level can be provided.

- Automatic manifolds are used in standard installations with maximum pressure requirements that do not exceed 1,000 kPa. For higher pressures, it is recommended that the manufacturer is contacted to confirm what modification is required, is available and is compatible with the pressures specified and that the manifold will meet the expected system demand.
- Automatic changeover manifolds are used where continuity of supply is mandatory.

5.7.8.2   MANUAL MANIFOLDS

Manual manifolds provide a continuous supply of gas that is connected to two independent banks of gas cylinders. Each bank is manually isolated and provides a continuous gas supply to the laboratory to allow for a replacement of the in-use cylinder bank while operating from the reserve supply. This will need to take into consideration the time necessary to purchase the replacement cylinders and to have them delivered to the laboratory ready for use. The changeover from the in use to the reserve cylinders is a manual operation and requires visual inspection of the gas supply on a regular basis.

- Used where monitoring of the gas supply is not critical or can be manually operated and visually monitored. These manifolds require a manual operation to alternate the supply side of the manifold.

5.7.8.3   SINGLE CYLINDER INSTALLATIONS

Small installations that do not require large volumes of gas may also be used for systems that can be manually operated and visually monitored.

- Used when certified or gravimetric gases are required and cost can be a consideration.

5.7.8.4   CYLINDER PACKS

Cylinder supply is the common method of providing systems with laboratory gases. They are available in cylinder packs containing up to fifteen cylinders that are interconnected and use a single or dual connection. The manifold and associated equipment used is similar to that used for other laboratory gases. However, cylinder valve connections may be gas specific; so it is necessary to specify the cylinder valve CGA number [20] and the number of cylinders that are required.

When using large numbers of cylinders it is necessary to specify the pressure and maximum flow rate for the pressure regulators to meet the system design flow rate.

5.7.8.5   PRESSURE SWING ADSORPTION (PSA) SYSTEMS

PSA systems that deliver nitrogen in a number of purity levels from 94% to 99.999% are available. They are obtainable in small bench top models that can provide sufficient gas for single gas chromatographs to large plant systems suitable for laboratory supply. The cost of the plant is relative to the purity and flow required, however, facilities that are unable to source regular nitrogen supplies may find this an attractive option.

The manufacturers should be contacted for information on PSA systems; they are supplied as complete plants and installed on site. The information required for the suppliers includes the required purity, supply pressure, and flow requirements. It is possible to provide these systems with backup manifolds that can be filled from local high-pressure compressors or from gas cylinder suppliers using manifold cylinders.

5.7.8.6   DUAL SUPPLY FOR CRYOGENIC SERVICES

Liquid and gaseous nitrogen are available for those with high demand systems; cryogenic vessels are supplied in capacities ranging from 3,000 to 25,000 liters from the majority of the gas suppliers. In most instances, the vessels are leased to laboratories under a contract between the gas manufacturer and the laboratory. A cryogenic vessel is a self-contained unit that includes all controls and pressure monitoring equipment as well as a heat exchanger for conversion of the liquid to its gas phase. Bulk liquid nitrogen supply would be used where high flows for extended applications are required or where long distances between the manufacturer's source of supply to the location of the laboratory or facility are significant.

5.7.8.7   ADDITIONAL SERVICES REQUIRED FOR CRYOGENIC VESSEL INSTALLATIONS

The use of bulk vessels requires additional services and supplies including:

- 240V power supply. Or as per national or local services.
- 415V 32-amp power supply. Or as per national or local services.
- Lighting.
- Security fencing.
- Water connection.
- Road access for a semi-trailer for delivery of liquid supplies.
- Reinforced concrete slab to be located as near to the road as possible to enable the liquid nitrogen delivery tanker easy access for filling purposes. The reinforced concrete slab for the support of the vessel will need to be designed by a structural engineer due to the weight of the vessel when full. The gas manufacturer may be able to assist with this.

- Alarm wiring or telemetry connections from the vessel control to a local monitoring panel may be required to monitor the vessel contents. This may also be connected to the gas supplier's depot for automatic ordering if it is available.
- Compliance with the National or Regional Authorities with suitable licenses obtained where applicable.
- Pressure regulation control panels may be supplied by the gas manufacturer at the source. However, that may not be the case in all locations and this equipment may need to be provided by the installation subcontractor. If this is the case a fully detailed specification in the design document will be necessary.

5.7.9   OUTLET VALVE TYPES/USES

- Laboratory taps can be either diaphragm (stainless steel diaphragm) or needle type suitably cleaned to the purity level of the gas supply.
- Double olive compression fittings are the most common connection type fitted to the outlets of the taps. Push on or serrated barb connections are unsuitable for the pressures encountered for the laboratory use of this gas.
- Valve manufacturers should be requested to confirm their range of purity levels available for the standard range of sizes.
- Needle type valves with double olive compression fittings are common where high and low-pressure connections are required.
- It is necessary to specify that the materials of construction including all components, seals, and seats are compatible with the gases being piped.
- The inclusion of laboratory taps of unknown cleanliness or purity and material compatibility should be avoided at all costs.
- Medical gas outlet connections are not suitable for laboratory service.

5.7.10   PURITY LEVEL

- Ultra-high purity: 99.999%.

5.7.11   ADDITIONAL DESIGN ASSISTANCE REFERENCES

- Adiabatic compression should be avoided wherever possible; the use of slow opening valves is recommended. Gay Lussac's law states that the pressure of a given amount of gas held at constant volume is directly proportional to the temperature in degrees Kelvin. When calculating the physical characteristics of any gas this may be of importance particularly where oxidizing, high pressure, toxic, non-respirable, and flammable gases are being installed.

- If solenoid valves are necessary the use of slow opening valve types is recommended. These can be controlled from local manually operated emergency push buttons and also interconnected to the fire trip system or any localized emergency gas monitoring system installed in the laboratory.

- Controlled pre-pressurized systems are an optional method of preventing adiabatic compression that can be caused by sudden activation of quick acting valves such as solenoid and ball valves.

- ASTM produces a number of documents that provide technical assistance for the design of gas systems. The ASTM G series publications [14–18] are of particular assistance for these types of installations.

- International and National codes of practice and regulations governing the use of dangerous goods must be referenced. In some countries, these regulations are protected by law and may not be ignored.

5.8   NITROGEN: CRYOGENIC LIQUID

5.8.1   DESCRIPTION

Gaseous nitrogen represents 78% of the earth's atmosphere; liquid nitrogen is distilled from the liquefied air during the manufacture of cryogenic liquids. It is supplied as a cryogenic liquid at a temperature of −196°C at pressures up to 1,000 kPa. It is a colorless, tasteless, odorless inert gas, and is an asphyxiant. It is non-flammable.

5.8.2 HAZARDS

Liquid nitrogen vaporizes to its gaseous state rapidly and creates additional hazards over and above those generated by the gaseous nitrogen. In its liquid state, it has problems that require further precautionary measures. The major concerns include:

- The temperature of the liquid is –196°C which will cause injury in its liquid state and frostbite if in a gaseous state should it come into contact with your skin.
- The liquid has an expansion ratio of liquid to gas of 682 to 1, i.e., one liter equals 682 liters of gas which at the boil off point will have the same temperature as the liquid, i.e., –196°C. The liquid to gas expansion rate is extreme and in an enclosed space the volume that has evaporated liquid nitrogen trapped will increase in pressure sufficiently to rupture the majority of commercially available pipeline materials and equipment.
- The gas manufacturers will provide the information necessary as well as advice on the National or Regional Authorities who have jurisdiction over the installation of cryogenic vessels as well as the precautions that are necessary when installing pipelines for cryogenic liquids.
- Gaseous nitrogen is an asphyxiant that can cause nausea and vomiting followed by a loss of consciousness and death. Loss of consciousness may occur without warning in areas of oxygen concentration below 19%. Personnel may not be physically aware of the effects of the gas and may become unconscious without warning. It is highly recommended that locations using this gas have permanent gas monitoring systems with local and remote audible and visual alarms fitted, the provision of warning signs in the area where flammable, toxic, and non-respirable gases are installed is recommended.
- Gaseous nitrogen is a compressed gas that can cause physical injury to the user or other nearby individuals. An uncontrolled pressurized gas flow provides a source of energy that can accelerate particles and other small items to velocities that may cause physical injury to the user or other nearby individuals.

5.8.3 USES

- Liquid nitrogen is used as a cryogenic freezing agent for the long-term storage of live cells, bacteria, sperm, plant cells, and similar applications. It can be connected to vacuum insulated storage vessels (cryovacs) for storage of cells using an automatic refill as the liquid level falls due to evaporation or is used for manually filling smaller local dewars.
- It is used as a cooling medium for Cryostats and in Molecular Beam Epitaxy as examples.

5.8.4 REFERENCES

Material Safety Data Sheets are available from the gas manufacturers that provide detailed information on this liquid. The information required for any laboratory gas design must include all of the relevant properties of the gas, any compatibility issues that are applicable to the design and the acceptable pressure, flow, concentration, and other pertinent factors that are applicable to the design [19, 21, 22, 24, 28, 31, 33, 58, 60–62].

For cryogenic services, it is also necessary to contact National or Regional Authorities to obtain approvals or certification on the installation that is proposed.

Any lubricants must be compatible with liquid nitrogen at −196°C. Care must be taken to ensure that any selected lubricants that are volatile and may become airborne, will not be frozen in the pipeline and foul moving parts or contaminate the gas purity.

Seats and seals of valves and controls must be suitable for operation at cryogenic temperatures when working with liquid nitrogen and other cryogenic liquids. It should be taken into consideration that the properties of some materials will alter when subjected to cryogenic temperatures.

5.8.5 MATERIALS OF CONSTRUCTION

5.8.5.1 COPPER TUBE

Copper pipelines may be used with polyurethane insulation; the disadvantage with this type of installation is its relatively poor insulation properties and the physical dimensions of the insulation that may need to have a 100 mm or greater wall thickness depending on the diameter of the inner pipeline.

The polyurethane insulated pipeline will continuously vaporize greater volumes of liquid nitrogen due to heat in a leak through the insulation than the vacuum jacketed pipe. The gas vaporized within the tube can force the liquid nitrogen back to the storage vessel when not in use. If this occurs the pipeline will require a period of precooling before liquid nitrogen will again be available at the outlet when next used.

Alternatively, a gas/liquid phase separator installed as near to the point of use as possible will maintain the level of liquid in the pipeline, however, the continuous loss of gas (24/7) would be costly. The exhaust from the phase separator must be insulated for its entire length before it is exhausted from the building, this will cause a build-up of ice when in operation that can become quite large. Some cryogenic pipeline manufacturers provide vacuum jacketed stainless steel pipeline sections with electrically operated heaters to prevent this.

The tube must be delivered to site pre-cleaned, dehydrated, and capped ready for use. During construction, the tube must be kept off the floor and in a location that prevents contamination from particulates or fluids that may occur on the job site.

5.8.5.2   STAINLESS STEEL

Vacuum jacketed stainless steel pipelines are the most efficient method of piping liquid nitrogen. The drawback is the cost of the pipeline. These systems are provided with gas/liquid phase separators which will continuously bleed the evaporated gas due to the inefficiency of the insulation however due to the better effectiveness of the pipeline it will not be as costly. The cost is something that must be taken into consideration. If a phase separator is used the exhaust tube must be vented to atmosphere and should have an electronic heater fitted to prevent frosting and ice build-up at the end of the pipeline. The exhaust pipeline must also be insulated to prevent ice build-up along its length. If a heater is not used the build-up of ice at the end of the exhaust pipeline may be considerable.

The prefabricated sections of stainless steel cryogenic tube must be delivered to site pre-cleaned, dehydrated, and capped ready for use. Care must be taken to protect the tube as damaged sections will require to remanufacture at the supplier's facility. During construction, the tube must be kept off the floor, protected, and in a location that prevents contamination from particulates or fluids that may occur on the job site.

5.8.5.3   INSULATION

All pipeline insulation types will have a percentage inefficiency that increases with the diameter and type of the tube as well as the type of insulation being used.

- Vacuum jacketed stainless steel is the most efficient (and most expensive) and is available in two formats, static, and dynamic:
  - Static vacuum jacketed pipelines are evacuated and sealed at the time of manufacture.
  - Dynamic vacuum jacketed pipework is connected to a high vacuum pump that operates continuously.
- Polyurethane insulation is the cheaper alternative (polystyrene is not suitable due to its minimum working temperature). The design should be based on the efficiency required, the distance from the bulk vessel to the outlet and the cost factor of lost gas during periods of use versus the cost of continuous bleed off of waste gas through a phase separator.

5.8.5.4   BRASS

Used for the isolation, pressure regulation and control valves where copper pipelines are installed. The isolation valves have extended spindles that allow them to be encapsulated into the polyurethane insulation during construction which leaves the valve handle accessible.

Brass valves are not the most common type used in vacuum jacketed pipelines with the exception of the isolation valves used on the cryogenic vessels. The connections between the vessel and the pipeline will need to be insulated with polyurethane.

Note: Polystyrene is not suitable due to its minimum working temperature for use with liquid nitrogen pipelines or equipment due to its minimum acceptable working temperature.

5.8.5.5   SYNTHETIC TUBE

The synthetic tube is not suitable for use with cryogenic liquids.

5.8.6   PRESSURE REQUIREMENTS

The liquid nitrogen is not pressured adjustable after the liquid leaves the bulk vessel, if it is imperative that a lower pressure is to be provided this must be arranged with the gas manufacturer who may be able to adjust the outlet pressure of the cryogenic vessel to suit the application. It may also be adjusted at the phase separator if the unit has that function built in.

Note: **Safety issue** – the expansion of liquid nitrogen into its gaseous state results in an approx. 682 to 1 [1] increase in volume, all pipelines containing liquid nitrogen must have safety relief valves installed at any location where liquid nitrogen may be trapped, e.g., between two valves or regulators.

5.8.7   CRYOGENIC SUPPLY

Liquid gaseous nitrogen, a cryogenic gas, is available for high demand systems; cryogenic vessels are available in capacities ranging from 3,000 to 25,000 liters from the majority of the gas suppliers. In most instances, the vessels are leased to laboratories under a contract between the gas manufacturer and the laboratory. A cryogenic vessel is a self-contained unit that includes all controls and pressure monitoring equipment as well as a heat exchanger for conversion of the liquid to its gas phase. Bulk liquid nitrogen supply would be used where high flows for extended applications are required or where the distance between the manufacturer's source of supply to the location of the laboratory or facility is significant.

5.8.7.1   ADDITIONAL SERVICES REQUIRED FOR CRYOGENIC VESSEL INSTALLATIONS

The use of bulk vessels requires additional services and supplies including:

- 240V power supply. Or as per national or local services.
- 415V 32-amp power supply. Or as per national or local services.

- Lighting.
- Security fencing.
- Water connection.
- Road access for a semi-trailer for delivery of liquid supplies.
- Reinforced concrete slab to be located as near to the road as possible to enable the liquid nitrogen delivery tanker easy access for filling purposes. The reinforced concrete slab for the support of the vessel will need to be designed by a structural engineer due to the weight of the vessel when full. The gas manufacturer may be able to assist with this.
- Alarm wiring or telemetry connections from the vessel control to a local monitoring panel may be required to monitor the vessel contents. This may also be connected to the gas supplier's depot for automatic ordering if it is available.
- Compliance with the National or Regional Authorities with suitable licenses obtained where applicable.
- Pressure regulation control panels may be supplied by the gas manufacturer at the source. However, that may not be the case in all locations and this equipment may need to be provided by the installation subcontractor. If this is the case a fully detailed specification in the design document will be necessary.

5.8.8   CRYOGENIC PIPELINE CONSTRUCTION

Cryogenic pipelines are constructed using three methods, polyurethane insulated copper pipelines, static vacuum jacketed stainless steel, and dynamic vacuum jacketed stainless steel.

5.8.8.1   POLYURETHANE INSULATED PIPELINE

Used for industrial applications and small LN2 systems where heat in leak and vaporization within the pipeline is not a concern. These pipelines are more suited where the pipeline itself is only used on an ad hoc basis and for intermittent filling of portable dewars or some industrial situations where gasification of the liquid nitrogen within the pipeline will not create problems for the end user. Examples would be industrial freezing where the pipeline is continuously in use or local hand filling of portable cryogenic dewars requiring short-term usage.

This type of installation may use a PVC pipeline with an internal copper pipe supported internally; the PVC pipe is then filled with polyurethane on completion. It is the easiest and most economically viable type of installation; however, has the lowest insulation properties. It is also the easiest type of installation that allows the use of pipe and fittings easily available from most pipeline supply wholesalers.

Where the insulation of the pipeline is manufactured from polyurethane and the pipeline is copper there is a range of extended spindle cryogenic valves available that provide mechanical screwed connections. These allow the pipeline and valve body to be encased in the polyurethane casing as a single unit. The majority of valve manufacturers provide valves designed for this service.

5.8.8.2   STATIC VACUUM JACKETED PIPELINES

Static vacuum jacketed tube is used where LN2 is required at the terminal outlet on demand and is suitable for installations where multiple fixed outlets from a single main pipeline are installed. This type of pipeline may also be used in systems that require a continuous liquid nitrogen supply that has phase separators installed to remove gaseous nitrogen from the pipeline prior to application. The pipeline may be constructed using fixed premanufactured pipe sections that are assembled onsite, they require considerable coordination and exact engineering design to ensure that assembly in-situ is possible.

These systems may use flexible connections between the phase separator and the equipment using the liquid nitrogen. Sections of dynamic vacuum jacketed pipework are often used for this section of the pipework to deliver the LN2 to the equipment and to extract the waste gaseous nitrogen through the phase separator.

The use of vacuum jacketed valves and controls are highly recommended for this type of pipeline.

5.8.8.3   DYNAMIC VACUUM JACKETED PIPELINES

Dynamic vacuum jacketed tube is used where LN2 is required at the terminal outlet on demand and prefabricated pipework is not a suitable alternative and flexibility of the final outlet location is necessary.

This type of system may not provide the same high level of insulation, the outer interstitial space in the tubing is connected to a continuously operating high vacuum pump to provide the vacuum insulation. There may be a cost implication for this type of pipework as it may also be used to return the vaporized liquid nitrogen to the phase separator.

5.8.8.4   VALVES AND CONTROLS

Vacuum jacketed cryogenic valves are used where isolation of the pipeline is required for vacuum and dynamic jacketed pipework; they are manufactured with extended spindles to allow the valve handles to protrude through the vacuum jacketed pipework and form an integral part of the pipeline.

Pneumatically controlled valves are available for remote operation of the systems; advice should be sought from the manufacturers of the vacuum jacketed pipe for additional information. These valves are available from the manufacturers of the cryogenic pipelines.

5.8.8.5   EXHAUST LN2

Liquid nitrogen pipework must have pressure relief valves fitted to any section of the pipeline where liquid nitrogen may be trapped between any two points such as isolation valves, plant, and equipment or any space where the expanded liquid cannot be vented freely.

It is advisable to vent any relief valve exhaust to a location external to the building where it can diffuse freely into the atmosphere. He vented gas may be at the same pipeline temperature or close to it at the point of release and must be clear of any location where it may come into contact with staff or the public. The exhaust pipeline should be manufactured from the same material as the supply pipeline; it is also recommended that suitable exhaust heaters are fitted to this exhaust to prevent the build-up of ice at the termination of the pipeline.

Comment: It is recommended that any LN2 pipelines be designed by experienced engineers and assistance from the proposed equipment manufacturers of the pipeline. This is a very specialized pipeline that requires knowledge and experience that is available from the gas suppliers and equipment manufacturers.

5.8.9   OUTLET VALVE TYPES

- Usual practice is to connect to the laboratory equipment using mechanical connections, the termination of the pipeline would be close by and the final connection would be a vacuum jacketed stainless steel flexible hose.
- For dewar filling, an isolation valve adjacent to the cryogenic vessel would be appropriate with a suitable length of vacuum insulated stainless steel flexible hose connected to a dispenser for hand filling the dewar.
- Automatic dispensing systems are often fitted to cryostats that provide a continuous supply and provide additional safety for the personnel responsible for this process if required.

5.8.10   PURITY LEVEL

- Liquid nitrogen is available from some gas suppliers in UHP grade, 99.999%.
- Liquid nitrogen standard purity is acceptable for cryogenic freezing and similar purposes; if the liquid is being used for other purposes, e.g., dual use with separate gas and liquid withdrawal facility, it must be supplied at the same purity level required for the gaseous system, i.e., High Purity or Ultra-High Purity as necessary.

5.8.11   ADDITIONAL DESIGN ASSISTANCE REFERENCES

- The suppliers of the cryogenic liquid must be contacted to determine the system pressures available, if the system is being connected to an existing supply it is necessary to ensure that the pressures required by the extension are compatible with this existing supply. The architect or the laboratory manager should be consulted to determine the system pressure requirements as the storage vessel may need to be configured prior to delivery to site.
- Contact with the manufacturers of the vacuum jacketed pipelines is necessary to provide sufficient installation information to allow the work to proceed. Liquid nitrogen pipelines are pre-manufactured

off-site to suit the building or site, alterations to sections of the tube after manufacture is not possible.

- National Standards, if available, may provide information about the location and installation of cryogenic storage vessels and will be required to determine what regulations apply to the site of the vessel and any safety and protection facilities for the users and public.
- Dangerous goods regulations either National or International may contain a number of specific requirements relating to the installation of cryogenic vessels, these documents should be consulted as they may be enacted in National statutes and as such cannot be ignored. Any breach of any of these regulations may prevent the gas manufacturer from filling the vessel until the installation complies with those requirements.
- If solenoid valves are necessary the use of slow opening valve types is recommended. These can be controlled from local manually operated emergency push buttons and also interconnected to the fire trip system or any localized emergency gas monitoring system installed in the laboratory.\
- ASTM produces a number of documents that provide technical assistance for the design of gas systems; in particular, the ASTM G series publications [14–18] are of assistance for these installations.

5.9 NITROUS OXIDE

5.9.1 DESCRIPTION

A mildly sweet-smelling, non-flammable oxidizing agent, it is a colorless gas and is an asphyxiant. It is used as a mild anesthetic when mixed with oxygen in medical practice. It is most commonly called laughing gas and occurs naturally in the atmosphere. The gas has a damaging effect on the atmosphere as it is classified as a greenhouse gas.

Nitrous Oxide is supplied in cylinders at a pressure of approximately 5,238 kPa at which pressure the gas is in a liquid state. It is stored in the cylinder as a liquid and when supplied in steel cylinders it weighs approximately 94 kgs whereas a similar size cylinder of nitrogen will weigh approximately 68 kgs in its gaseous state. A gas cylinder containing nitrous oxide contains approximately 2.4 times the volume of gas when compared to a cylinder of UHP nitrogen. For UHP nitrogen this represents approximately 7,000liters of gas whereas a cylinder of nitrous oxide of the same size contains 17,000 liters of gas.

5.9.2 HAZARDS

- Nitrous oxide is an oxidizing agent that will vigorously support combustion, it has similar properties to oxygen and if in direct contact with combustible substances may cause them to ignite.
- Nitrous oxide is a compressed gas that can cause physical injury to the user or other nearby individuals. An uncontrolled pressurized gas flow provides a source of energy that can accelerate particles and other small items to velocities that may cause physical injury to the user or other nearby individuals.
- Nitrous oxide is an asphyxiant. To determine the levels for STEL and TWA contact should be made with National or Regional Authorities.
- Nitrous oxide has aesthetic properties that may cause disorientation, euphoria, and eventual unconsciousness.
- It is highly recommended that locations using this gas have permanent gas monitoring systems with local and remote audible and visual alarms fitted, the provision of warning signs in the area where flammable, toxic, and non-respirable gases are installed is recommended.
- The provision of warning signs in the area where non-respirable gases are installed is recommended.
- The use of nitrous oxide specific gas sensors is advisable due to the minimum percentage of nitrous oxide in the atmosphere that is considered harmful and any increase in concentration would be insufficient to initiate a low-level oxygen alarm.

5.9.3 LABORATORY USE

- Nitrous oxide is used in conjunction with acetylene for producing the flame used in AAS [23].
- Nitrous oxide is used in calibration gas mixtures for the petrochemical industry; environmental emission monitoring, industrial hygiene monitors, and trace impurity analyzers.

5.9.4 REFERENCES

Material Safety Data Sheets are available from the gas manufacturers that provide detailed information on this gas. The information required for any laboratory gas design must include all of the relevant properties of the gas, any compatibility issues that are applicable to the design and the acceptable pressure, flow, concentration, and other pertinent factors that are applicable to the design [19, 21, 22, 24, 28, 31, 33, 58, 60–62].

Many lubricants are compatible with nitrous oxide, however, equipment manufacturers must be contacted to confirm their equipment is suitable for purpose and care must be taken to ensure that any selected volatile lubricants that may vaporize in the gas stream will not contaminate the gas purity.

Some seat and seal materials are compatible at low pressures however at full cylinder pressure this is not so, care must be taken to ensure any materials are suitable for use at cylinder pressure when designing high-pressure systems.

Some synthetic hose materials are not suitable for long-term exposure to high or low-pressure nitrous oxide. It has been shown that this gas has the ability to extract plasticizers from some synthetic hose types.

5.9.5 MATERIALS OF CONSTRUCTION

5.9.5.1 COPPER TUBE

Copper tube for use with nitrous oxide should be as supplied for medical grade gases in an oxygen clean state is suitable for the majority of nitrous oxide gas pipelines [19, 61]. The tube must be delivered to site pre-cleaned, dehydrated, and capped ready for use. During construction, the tube must be kept off the floor and in a location that prevents contamination from particulates or fluids that may occur on the job site. Medical Gas Grade copper tube is used in the majority of laboratory gas pipelines [19, 61]. All pipe used in laboratory gas service must be cleaned for oxygen service.

It is recommended that hard drawn or semi-hard copper tube is used for laboratory gas pipelines unless there are incompatibility issues or it is recommended otherwise. The use of soft annealed copper pipe is not recommended for laboratory gas pipelines.

5.9.5.2　STAINLESS STEEL

Used for special instances; however, there is no advantage in the use of stainless steel unless there is a specific compatibility or cleanliness issue that may corrupt the gas supply. The pipeline design must not impact the gas in any way and thereby ensure the gas purity levels are unchanged when delivered at the outlet connection. Stainless steel is recommended for high-pressure service where the copper tube is not suitable due to pressure limitations or availability however special design considerations must be used to allow for the properties of this gas.

5.9.5.3　ELECTROPOLISHED STAINLESS STEEL TUBE

The electropolished stainless steel tube is produced specifically for the electronics industry. It should be noted that this tube should have the same cleanliness level as certified oxygen clean tube.

The electropolished stainless steel tube is used to assist in the prevention of contamination due to outgassing where spontaneously combustible gases are piped and gases may be trapped in the pipe surface.

Note: It has been suggested that the use of this tube type will improve the flow characteristics of a pipeline. However, calculations using the Darcy Weisbach formula [21, 22, 24, 58] to calculate critical pipeline flows indicate that any increase that may occur is of little or no consequence in a laboratory gas pipeline. Any increase in flow that may be provided in using this tube is less than 0.1% of the design flow rate, the use of oxygen cleaned unpolished stainless steel tube of the same size will not impact on the working pressure or design flow. The design flow and system pressure should use a design safety margin normally set at 10% above the system working parameters.

5.9.5.4　BRASS

Brass is used in the manufacture of isolation, and control valves. It is also the most common type used in the production of pressure regulators in the majority of cases for gas purity levels up to ultra-high purity at 99.999%. Some laboratory bench mounted tapware is manufactured from brass with an external protective coating.

All valves used for the control of laboratory gas pipelines must be oxygen cleaned equally to the national standard for medical grade gases as a minimum; certification of cleanliness levels should be provided by the manufacturer.

5.9.5.5   SYNTHETIC TUBE

The synthetic tube has been suggested for use in high and low-pressure applications. If the synthetic tube is used for high-pressure service it must be rated for the maximum working pressure of the cylinders or pipeline and provide safe working conditions for the laboratory staff. It is used in some high-pressure hoses for cylinder connections to manifolds; however, this tubing is reinforced with stainless steel outer braiding or similar protection.

For low-pressure applications synthetic tube has a different set of problems, these include:

- The emission of volatile organic compounds used in the manufacture of the tube.
- The emission of esters used in the manufacture of the tube.
- The material used in some synthetic tubing is hygroscopic.
- Synthetic tubing and fittings may not be available in an oxygen clean state.
- Synthetic tubing may be porous to some gases, helium, and mixtures containing it are subject to this effect.

Note: The use of synthetic tube is not recommended for laboratory gas pipelines, the tubing may have materials concerns that could contaminate the gas stream, if specifically required the suppliers should provide confirmation that the material is fit for purpose.

5.9.6   PRESSURE REQUIREMENTS

For the majority of laboratory nitrous oxide systems, a pipeline pressure of 410 kPa will be sufficient. It is highly recommended that consultation with the laboratory staff is undertaken to ensure that the gas being supplied will meet with their requirements and the equipment being used. There are continually new procedures and equipment that may require gases provided in a unique manner that is not able to be met by a standard laboratory gas pipeline system.

The working pressure of the pipeline will determine what types and models of valves will need to be used as standard laboratory tapware may vary in operating pressures between 700 and 1,000 kPa. High-pressure systems may require specialized pipeline materials, valves, and tapware that will have to be individually specified to meet these requirements; advice from the manufacturers will be necessary for this type of installation and the design criteria will need to be detailed in any specification to provide information on the proposed working environment.

Note: Safety issue – the expansion of liquid nitrous oxide into its gaseous state results in a 662 to 1 increase in volume [1] this conversion rate may vary between Material Safety Data Sheets depending on specified temperature. All pipelines containing high-pressure liquid nitrous oxide must have safety relief valves with the relief port vented to atmosphere and installed at any location where liquid nitrous oxide may be trapped, e.g., between two valves or controls.

5.9.7   GAS SUPPLY REQUISITES

There are a number of options available for the supply of gases in laboratories. The factors that determine which type of supply source will be optimal will depend on a number of elements:

5.9.7.1   SYSTEM FLOW DEMAND

Every system will need to be designed to suit the equipment being used in the laboratory, e.g., a small single laboratory using one bench mounted AAS [23] may require a nitrous oxide supply that may be supplied from a single cylinder that can be monitored on a daily basis. A major laboratory using a number of AASs [23] would require considerably higher flows and it is imperative to have information on the types and flow demand of the various items of equipment prior to determining the gas supply capacity.

For nitrous oxide, the flow rate may vary between laboratories as the uses range from low flows for individual locations of 20 liters per minute to central facilities that require flows to meet the demand of each of the AASs [23] operating concurrently at 20 liters per minute.

5.9.7.2   SYSTEM PRESSURE

The high-pressure liquid nitrous oxide is supplied at a cylinder pressure of 5,824 kPa; it is not usually piped at high pressure. The cylinder pressure is constant until all of the liquid remaining in the cylinder has evaporated at which time the pressure will reduce rapidly until empty. The pipeline must be designed with this taken into consideration.

5.9.7.3   SYSTEM DEPENDABILITY

Each gas system will have requirements that dictate the type of supply required, nitrous oxide is no exception. Many laboratories require a permanent supply that will be online twenty-four hours a day and seven days a week. This can be accomplished for most gas supply systems using manifold cylinders, the system will need a service supply that will provide notification to laboratory staff that the system is in its standard operating condition or has a depleted gas supply caused by usage or an equipment failure.

To overcome this problem the cylinder manifold is provided with a backup or reserve supply. For a cylinder supply, the number of cylinders required for operation is duplicated with an alarm function to automatically advise staff that the change from the online supply to the reserve supply has occurred.

5.9.8   GAS SUPPLY SYSTEMS

5.9.8.1   AUTOMATIC MANIFOLDS

Automatic manifolds provide a continuous supply of gas that is connected to two independent banks of gas cylinders. Each bank should have the ability to automatically provide a continuous gas supply to the laboratory to allow for a replacement of the in-use cylinder bank while operating from the reserve supply. This will need to take into consideration the time necessary to purchase the replacement cylinders and to have them delivered to the laboratory ready for use.

Purity levels specified by the laboratory must be met by the manifold supplier to ensure compatibility with the gas being piped. Certification should be sought from the recommended supplier that the specified purity level can be provided.

- Automatic manifolds are used in standard installations with recommended pressure requirements of 410 kPa.
- Automatic changeover manifolds are used where continuity of supply is mandatory.

5.9.8.2   MANUAL MANIFOLDS

Manual manifolds provide a continuous supply of gas that is connected to two independent banks of gas cylinders. Each bank is manually isolated and provides a continuous gas supply to the laboratory to allow for a replacement of the in-use cylinder bank while operating from the reserve supply. This will need to take into consideration the time necessary to purchase the replacement cylinders and to have them delivered to the laboratory ready for use. The changeover from the in use to the reserve cylinders is a manual operation and requires visual inspection of the gas supply on a regular basis.

- Used where monitoring of the gas supply is not critical or can be manually operated and visually monitored. These manifolds require a manual operation to alternate the supply side of the manifold.

5.9.8.3   SINGLE CYLINDER INSTALLATIONS

Small installations that do not require large volumes of gas may also be used for systems that can be manually operated and visually monitored.

5.9.9   BULK SUPPLY ALTERNATIVES

5.9.9.1   CYLINDER PACKS

Cylinder supply is the common method of providing systems with laboratory gases. They are available in cylinder packs containing up to 15 cylinders although this number may be less for nitrous oxide cylinders, they are interconnected and use a single or dual connection. The manifold and associated equipment used is similar to that used for other laboratory gases. However, cylinder valve connections may be gas specific; so it is necessary to specify the cylinder valve CGA number [20] and the number of cylinders that are required.

When using large numbers of cylinders it is necessary to specify the use of pressure regulators that are able to meet the system design flow rates.

5.9.10   OUTLET VALVE TYPES/USES

- Laboratory taps can be either diaphragm (stainless steel) or needle type and must be suitably cleaned to meet the purity level of the gas supply. Taps with double olive compression fittings are the most common connection type used as the outlet for the outlets of the taps.
- For all valve manufacturers, it is imperative to check their range of variable purity levels for availability, if there is no available information it may indicate that the supplier in question is unable to provide conforming products.
- It is recommended that slow opening needle type valves are used with double olive compression fittings. This is particularly important where high-pressure gases are piped.
- The inclusion of laboratory taps of unknown cleanliness or purity and material compatibility should be avoided at all costs.
- Medical gas outlet connections are not suitable for laboratory service.

5.9.11   PURITY LEVEL

- Ultra-high purity: 99.999%.

5.9.12   ADDITIONAL DESIGN ASSISTANCE REFERENCES

- Flashback arrestors should be fitted to nitrous oxide pipelines when connected to an AAS [23], they are available in a number of formats and sizes and must be selected to suit the application taking flows and flash arrestor types into consideration. There are some units available that include a mechanical non-return valve that may create a noise as the valve opens and closes. This noise can register in the AAS [23] and cause irregularities in the printout, this type is not suitable for laboratory use.

- Adiabatic compression should be avoided wherever possible; the use of slow opening valves is recommended. Gay Lussac's law states that the pressure of a given amount of gas held at constant volume is directly proportional to the temperature in degrees Kelvin. When calculating the physical characteristics of any gas this may be of importance particularly where oxidizing, high pressure, toxic, non-respirable, and flammable gases are being installed.
- If solenoid valves are necessary the use of slow opening valve types is recommended. These can be controlled from local manually operated emergency push buttons and also interconnected to the fire trip system or any localized emergency gas monitoring system installed in the laboratory.
- Controlled pre-pressurized systems are an optional method of preventing adiabatic compression that can be caused by sudden activation of quick acting valves such as solenoid and ball valves.
- ASTM produces a number of documents that provide technical assistance for the design of gas systems. The ASTM G series publications [14–18] are of particular assistance for these types of installations.

5.10   OXYGEN

5.10.1   DESCRIPTION

Gaseous oxygen is a tasteless, colorless, and odorless gas comprising approximately 21% of the earth's atmosphere. It is non-flammable; however, it will vigorously support combustion at levels above 21% in the air, it has the capability to change some materials into combustible products that would not otherwise burn in air. It is an oxidizing agent and fire accelerant.

5.10.2   HAZARDS

- Many substances will ignite in enriched oxygen atmospheres (recognized values of >26% for materials and >24% for medical purposes) that will not ignite in air. In laboratory use, these substances include valve seats and sealing materials as well as the valves and pipeline materials themselves. Under certain conditions, stainless steel will act as a fuel source when in an oxygen-enriched atmosphere. The use of any hydrocarbon lubricants is to be avoided at all costs, specific oxygen compatible lubricants are available however these must be checked for suitability with respect to pressure and temperature for the conditions that are to be encountered as well as compatibility with the gas purity.

- There are three requirements to create a fire, oxygen, a fuel source, and heat, a combination of these three in varying amounts will cause an ignition. The oxygen is a variable that must be present for any fire however its effect on the other two can be dramatic, increased oxygen levels permit lower temperature ignition or provide conditions where materials will burn that otherwise would not combust in air. From a gas, perspective heat can be supplied externally or by the gas itself, e.g., by adiabatic compression that can easily supply a heat source and in enriched oxygen atmospheres combustion may occur with explosive force at lower temperatures than would occur otherwise.

- Gaseous oxygen has been known to be absorbed into clothing when working in areas where the atmospheric oxygen concentration has been increased, this can make clothing a fuel source and should any heat source no matter how small come into contact with the other two it will ignite instantly with catastrophic results.

- It is highly recommended that locations using this gas have permanent gas monitoring systems with local and remote audible and visual alarms fitted, the provision of warning signs in the area where flammable, toxic, and non-respirable gases are installed is recommended.

5.10.3   USES

Oxygen is used in metallurgical processes, generation of high-temperature flames and in oxygen bomb calorimeters. The most common laboratory use is for the measurement of the calorific value of substances in an oxygen bomb.

5.10.4   COMPATIBILITY ISSUES

The selection of any equipment for use in oxygen service must take the particular properties of oxygen into consideration. When oxygen is encountered in an enriched state (>26% for materials and >24% for medical purposes) it provides conditions that allow otherwise non-flammable materials to become highly flammable. An increase of as little as 3% (from 21% to 24% oxygen in air) has been shown to change the properties of the air sufficiently that it alters the ability of some materials to burn. Materials that burn freely in the air will ignite with increased violence in an oxygen-enriched atmosphere, this may easily put lives in danger should they come in contact with an undetectable oxygen-enriched atmosphere. Hydrocarbons are of particular concern hence the warnings about maintaining a safe distance between flammable materials and oxygen sources.

The use of lubricating substances must be done with extreme care, hydrocarbon-based oils and greases will ignite freely in oxygen with explosive force. Some metals may provide a fuel source for oxygen and design procedures must be undertaken with extreme care.

5.10.5   REFERENCES

Material Safety Data Sheets are available from the gas manufacturers that provide detailed information on this gas. The information required for any laboratory gas design must include all of the relevant properties of the gas, any compatibility issues that are applicable to the design and the acceptable pressure, flow, concentration, and other pertinent factors that are applicable to the design [19, 21, 22, 24, 28, 31, 33, 58, 60–62].

A limited number of lubricants are compatible with oxygen; however, it is not recommended that these be used in parts of a pipeline that are in direct contact with the gas stream. If necessary to use lubricants the manufacturers must be contacted to confirm their products are suitable for the intended purpose and will not contaminate the purity of the oxygen. Care must be taken to ensure that any volatile lubricants or other contaminants that may enter the gas stream and contaminate the gas supply purity are specifically excluded.

5.10.6   MATERIALS OF CONSTRUCTION

5.10.6.1   COPPER TUBE

Copper tube for use with gaseous oxygen should be as supplied for medical grade gases in an oxygen clean state is suitable for the majority of oxygen gas pipelines [19, 61]. The tube must be delivered to site pre-cleaned, dehydrated, and capped ready for use. During construction, the tube must be kept off the floor and in a location that prevents contamination from particulates or fluids that may occur on the job site. Medical Gas Grade copper tube is used in the majority of laboratory gas pipelines [19, 61].

It is recommended that hard drawn or semi-hard copper tube is used for laboratory gas pipelines unless there are incompatibility issues or it is recommended otherwise. The use of soft annealed copper pipe is not recommended for laboratory gas pipelines.

5.10.6.2   STAINLESS STEEL

Used for special instances; however, there is no advantage in the use of stainless steel unless there is a specific compatibility or cleanliness issue that may corrupt the gas supply. The pipeline design must not impact the gas in any way and thereby ensure the gas purity levels are unchanged when delivered at the outlet connection point. Stainless steel is not recommended for high-pressure oxygen service as it may become a fuel source, if it is necessary to use stainless steel the designer must be fully conversant with the problems associated with the use of this material at high pressures.

5.10.6.3   ELECTROPOLISHED STAINLESS STEEL TUBE

The electropolished stainless steel tube is produced specifically for the electronics industry. It should be noted that this tube should have the same cleanliness level as certified oxygen clean tube.

The electropolished stainless steel tube is used to assist in the prevention of contamination due to outgassing where spontaneously combustible gases are piped and gases may be trapped in the pipe surface.

Note: It has been suggested that the use of this tube type will improve the flow characteristics of a pipeline. However, calculations using the

Darcy Weisbach formula [21, 22, 24, 58] to calculate critical pipeline flows indicate that any increase that may occur is of little or no consequence in a laboratory gas pipeline. Any increase in flow that may be provided in using this tube is less than 0.1% of the design flow rate, the use of oxygen cleaned unpolished stainless steel tube of the same size will not impact on the working pressure or design flow. The design flow and system pressure should use a design safety margin normally set at 10% above the system working parameters.

5.10.6.4   MONEL

Specifically used for high-pressure systems, it is not commonly found in laboratories.

5.10.6.5   BRASS

Brass is used in the manufacture of isolation, and control valves. It is also the most common type used in the production of pressure regulators, in the majority of cases for gas purity levels up to ultra-high purity at 99.999%. Some laboratory bench mounted tapware is manufactured from brass with an external protective coating.

All valves used for the control of laboratory gas pipelines must be oxygen cleaned equally to the national standard for medical grade gases as a minimum; certification of cleanliness levels should be provided by the manufacturer.

For high-pressure applications, it is necessary to develop specific design procedures that will ensure the compatibility of the tube and fittings with this gas. Materials such as stainless steel, mild steel may become fuel sources in high-pressure oxygen applications. Reference to the ASTM G series documents provides additional information on this subject.

5.10.6.6   SYNTHETIC TUBE

The synthetic tube has been suggested for use in high and low-pressure applications. If the synthetic tube is used for high-pressure service it must be rated for the maximum working pressure of the cylinders or pipeline and provide safe working conditions for the laboratory staff. It is used in some high-pressure hoses for cylinder connections to manifolds; however, this tubing is reinforced with stainless steel outer braiding or similar protection.

For low-pressure applications synthetic tube has a different set of problems, these include:

- The emission of volatile organic compounds used in the manufacture of the tube.
- The emission of esters used in the manufacture of the tube.
- The material used in some synthetic tubing is hygroscopic.
- Synthetic tubing and fittings may not be available in an oxygen clean state.
- Synthetic tubing may be porous to some gases, helium, and mixtures containing it are subject to this effect.

Note: The use of synthetic tube is not recommended for laboratory gas pipelines. The tubing may have materials concerns that could contaminate the gas stream, if specifically required the suppliers should provide confirmation that the material is fit for purpose.

5.10.7   PRESSURE REQUIREMENTS

Variable pressures are required depending on the application, e.g., oxygen bomb calorimeters may need in excess of 5,000 kPa depending on the model and type of calorimeter. It is important to determine which model or type of calorimeter the pipeline system is being designed to connect to.

5.10.8   GAS SUPPLY SYSTEMS

5.10.8.1   AUTOMATIC MANIFOLDS

Automatic manifolds [29] provide a continuous supply of gas that is connected to two independent banks of gas cylinders. Each bank should have the ability to automatically provide a continuous gas supply to the laboratory to allow for a replacement of the in-use cylinder bank while operating from the reserve supply. This will need to take into consideration the time necessary to purchase the replacement cylinders and to have them delivered to the laboratory ready for use.

Purity levels specified by the laboratory must be met by the manifold supplier to ensure compatibility with the gas being piped. Certification should be sought from the recommended supplier that the specified purity level can be provided.

- Automatic manifolds are used in standard installations with maximum pressure requirements that do not exceed 1,000 kPa. For higher pressures, it is recommended that the manufacturer is contacted to confirm what modification is required, is available and is compatible with the pressures specified and that the manifold will meet the expected system demand.
- Automatic changeover manifolds are used where continuity of supply is mandatory.

5.10.8.2   MANUAL MANIFOLDS

Manual manifolds provide a continuous supply of gas that is connected to two independent banks of gas cylinders. Each bank is manually isolated and provides a continuous gas supply to the laboratory to allow for a replacement of the in-use cylinder bank while operating from the reserve supply. This will need to take into consideration the time necessary to purchase the replacement cylinders and to have them delivered to the laboratory ready for use. The changeover from the in use to the reserve cylinders is a manual operation and requires visual inspection of the gas supply on a regular basis.

- Used where monitoring of the gas supply is not critical or can be manually operated and visually monitored. These manifolds require a manual operation to alternate the supply side of the manifold.

5.10.8.3   SINGLE CYLINDER INSTALLATIONS

Small installations that do not require large volumes of gas may also be used for systems that can be manually operated and visually monitored.

- Used when certified or gravimetric gases are required and cost can be a consideration.

5.10.9   BULK SUPPLY ALTERNATIVES

5.10.9.1   CYLINDER PACKS

Cylinder supply is the common method of providing systems with labora-tory gases. They are available in cylinder packs containing up to fifteen cylinders that are interconnected and use a single or dual connection. The manifold and associated equipment used is similar to that used for other laboratory gases. However, cylinder valve connections may be gas specific; so it is necessary to specify the cylinder valve CGA number [20] and the number of cylinders that are required.

When using large numbers of cylinders it is necessary to specify the pressure and maximum flow rate for the pressure regulators to meet the system design flow rate.

5.10.9.2   LIQUID OXYGEN SUPPLY

Liquid oxygen is a cryogenic gas and is available for high demand systems, cryogenic vessels are available in capacities ranging from 3,000 to 25,000 liters from the majority of the gas suppliers. In most instances, the vessels are leased to laboratories under a contract between the gas manufacturer and the laboratory. A cryogenic vessel is a self-contained unit that includes all controls and pressure monitoring equipment as well as a heat exchanger for conversion of the liquid to its gas phase. Bulk liquid oxygen supply is used where high flows for extended applications are required or extended distances between the manufacturer's source of supply to the location of the laboratory or facility are significant.

5.10.9.3   ADDITIONAL SERVICES REQUIRED FOR CRYOGENIC VESSEL INSTALLATIONS

The use of bulk vessels requires additional services and supplies including;

- 240V power supply. Or as per national or local services.
- 415V 32-amp power supply. Or as per national or local services.
- Lighting.
- Security fencing.

- Water connection.
- Road access for a semi-trailer for delivery of liquid supplies.
- Reinforced concrete slab to be located as near to the road as possible to enable the liquid oxygen delivery tanker easy access for filling purposes. The reinforced concrete slab for the support of the vessel will need to be designed by a structural engineer due to the weight of the vessel when full. The gas manufacturer may be able to assist with this.
- Alarm wiring or telemetry connections from the vessel control to a local monitoring panel may be required to monitor the vessel contents. This may also be connected to the gas supplier's depot for automatic ordering if it is available.
- Compliance with the National or Regional Authorities with suitable licenses obtained where applicable.
- Pressure regulation control panels may be supplied by the gas manufacturer at the source. However, that may not be the case in all locations and this equipment may need to be provided by the installation subcontractor. If this is the case a fully detailed specification in the design document will be necessary.

5.10.10   OUTLET VALVE TYPES/USES

- Laboratory taps can be either diaphragm (stainless steel diaphragm) or needle type suitably cleaned to the purity level of the gas supply, Double olive compression fittings are the most common connection type fitted to the outlets of the taps. Push on or serrated barb connections are unsuitable for the pressures encountered for the laboratory use of this gas.
- Valve manufacturers should be requested to confirm their range of purity levels available and the standard range of sizes.
- Needle type valves with double olive compression fittings are common where high and low-pressure connections are required. It is necessary to specify that the materials of construction including all components, seals, and seats are compatible with the gases being piped.
- The inclusion of laboratory taps of unknown cleanliness or purity and material compatibility should be avoided at all costs.

• Medical gas outlet connections are not suitable for laboratory service.

5.10.11   PURITY LEVELS

• Ultra-high purity: 99.999%.

Note: Medical grade gases are 99.5% pure and are not suitable for laboratory service.

5.10.12   ADDITIONAL DESIGN ASSISTANCE REFERENCES

• Adiabatic compression should be avoided wherever possible; the use of slow opening valves is recommended. Gay Lussac's law states that the pressure of a given amount of gas held at constant volume is directly proportional to the temperature in degrees Kelvin. When calculating the physical characteristics of any gas this may be of importance particularly where oxidizing, high pressure, toxic, non-respirable, and flammable gases are being installed.

• If solenoid valves are necessary the use of slow opening valve types is recommended. These can be controlled from local manually operated emergency push buttons and also interconnected to the fire trip system or any localized emergency gas monitoring system installed in the laboratory.

• Controlled pre-pressurized systems are an optional method of preventing adiabatic compression that can be caused by sudden activation of quick acting valves such as solenoid and ball valves.

• ASTM produces a number of documents that provide technical assistance for the design of gas systems. The ASTM G series publications are of particular assistance for these types of installations.

• International and National codes of practice and regulations governing the use of dangerous goods must be referenced. In some countries, these regulations are protected by law and may not be ignored.

• There is a maximum recommended velocity for piped gaseous oxygen of 250 mm/sec [Ref. ASTM].
  Note: At this velocity, there would be pressure losses that would be unacceptable in the majority of laboratory gas systems.

- As pressure levels increase so does the risk with piped oxygen, at any pressure in excess of 410 kPa oxygen-related design references must be used for the compatibility issues that will be encountered. ASTM produces a number of documents [14–18] that provide technical assistance for these issues.
- NASA produces technical papers on the use of high-pressure oxygen that are available for purchase if additional information is required.

5.11 VACUUM

5.11.1 DESCRIPTION

Vacuum can be described as a volume in which the internal pressure is below atmospheric pressure, the differential pressure between the two indicates the mass of gas that has been evacuated from the nominated volume.

For laboratory purposes the level of vacuum may be measured in either absolute or gauge pressures, Absolute (A) pressure indicates the pressure of any gas based on an absolute value, in other words zero kPa (A) is equal to the absence of the gas pressure usually provided by the atmosphere whereas Gauge pressure (G) represents the pressure from a zero starting point of atmospheric pressure. As a comparison zero kPa (A) is equal to –101.35 kPa (G) or alternatively zero kPa (G) is equal to plus 101.35 kPa (A).

For the majority of calculations when using vacuum many engineers use 100 kPa as the rough calculation starting point for atmospheric pressure however when working with the high vacuum it is necessary to use the correct pressure to at least two decimal places to ensure accuracy.

For the purposes of this book all pressures are measured at Gauge pressure, should Absolute pressure be used for clarity it has (A) appended to the value.

5.11.2 HAZARDS

- Contaminants that may be drawn into plant and pipelines can condense in pumps and cause premature wear and physical damage.
- Electrical control panels incorporate high voltage wiring and switchgear.

- Toxic or contaminated waste vapors may be found in the exhaust pipeline.
- Vacuum exhaust pipelines can cause burns through direct personal contact.
- Direct contact with evacuated vacuum pipelines and valves may cause rupture to surface blood vessels.
- If oil sealed vacuum pumps are used consideration of the operating pressure and the exhaust pipeline temperature must be calculated when selecting the location of the exhaust and any safety measures required to protect personnel from any physical exposure. Insulation of the exhaust pipeline is a recommended method of providing protection.

5.11.3   COMPATIBILITY ISSUES

There are compatibility concerns in vacuum systems that relate to the condensed products that are likely to be drawn into the system during use. There is a variety of aggressive VOCs that are used in laboratories that may corrode some materials used in the manufacture of plant and equipment. These must be investigated prior to the design stage and the plant selected accordingly.

5.11.4   USES

- Extraction of VOCs used to separate proteins and similar materials during laboratory procedures.
- Removal of unwanted liquids or contaminants.
- Removal of fumes and gases.

5.11.5   MATERIALS OF CONSTRUCTION

5.11.5.1   COPPER TUBE

Copper tube is used in the majority of vacuum systems, the exemption is where there is the probability that non compatible substances may be in contact with the pipeline. All pipe used in laboratory gas service must be cleaned for oxygen service. The supply of oxygen cleaned tube for vacuum service is not specifically for contamination issues, it is used to prevent the inadvertent use of uncleaned tube on a High Purity laboratory gas pipeline.

5.11.5.2   STAINLESS STEEL

Used for systems where aggressive products or other materials that may be drawn into the pipework during operation are incompatible with copper.

Note: The selection of vacuum pumps, valves, and pipeline fittings should mirror this selection.

5.11.5.3   BRASS

Brass is used in the manufacture of isolation, and control valves. It is also the most common type used in the production of pressure regulators which are rarely installed in laboratory vacuum pipelines. Some laboratory bench mounted tapware is manufactured from brass with an external protective coating.

All valves used for the control of laboratory gas pipelines must be oxygen cleaned equally to the national standard for medical grade gases as a minimum; certification of cleanliness levels should be provided by the manufacturer. This requirement is recommended for valves and pipes used in vacuum service to prevent the inadvertent inclusion of uncleaned equipment in the laboratory gas pipeline.

5.11.5.4   SYNTHETIC TUBE

If the synthetic tube is used for vacuum service it must be manufactured for use at vacuum pressures and may require increased wall thickness or internal reinforcement to prevent the pipe from collapsing when vacuum is applied.

The following are concerns that have been found in some synthetic pipeline systems:

- The collection of volatile organic compounds used in the laboratory that may adversely affect the synthetic pipe.
- Protection may be necessary to protect the synthetic tube from physical damage during use.
- Incompatibility with some VOCs and chemicals used in laboratories. **Note**: The use of a synthetic tube is not recommended for laboratory vacuum pipelines.

5.11.6 PRESSURE REQUIREMENTS

For the majority of laboratory uses a pipeline pressure of –60 kPa will suffice, it is highly recommended that consultation with the laboratory staff is undertaken to ensure that the vacuum level being supplied will meet with their requirements to suit the equipment and procedures proposed. There are always new procedures and equipment being introduced that may require vacuum service to be provided in a unique manner that is not met by standard systems.

There are some instances where high vacuum levels lower than –99 kPa are required, this application should be undertaken using localized single outlet vacuum pumps as the means of supply. These small independent systems are suitable for recovery of some volatile organic compounds and specialist localized pump types are available providing optional condensing coils that require connection to a chilled water supply.

For vacuum pressures in the range of –60 kPa to –85 kPa central systems that use a high volume vacuum pump plant would be necessary to evacuate the high volume of expanded gas being evacuated from the pipeline. As an example, a pump that is designed to remove 40 liters per minute of free air when operating at –60 kPa will need to extract approximately 100 liters per minute of rarefied air. If the pump is operating at –99 kPa when operating at the same flow rate it will need to extract 2,000 liters per minute of rarefied air. It should also be noted that the pumps used for high vacuum service are not suitable to operate in rough vacuum service and rough vacuum pumps cannot operate at high vacuum. It is absolutely necessary to calculate the pressure/flow requirements for any vacuum system and to select the correct types and models required for the service.

5.11.7   PLANT SUPPLY ALTERNATIVES

Refer to Chapter 2 for detailed information on plant types and uses.

5.11.8   OUTLET VALVE TYPES/USES

- Laboratory taps using large orifice diaphragm operation are the preferred option; push on barbed tails are most the common type fitted to the outlets of the taps in use at –60 kPa; however, for high vacuum service taps that have been specifically designed for high vacuum use should be selected.
- Specialist valve manufacturers should be contacted to confirm that the range of equipment available is suitability for the vacuum levels and their availability.
- Valves with integral compression fittings or face type couplings are common where high vacuums are encountered.
- Medical gas outlet connections are not suitable for laboratory vacuum service.

5.11.9   PRESSURE LEVELS/TYPES

- For general usage vacuum levels between –60 kPa to –85 kPa are suitable, this can be used for connection to rotary evaporators or similar that can have high flow demands at this vacuum level.
- Levels of vacuum between –85 kPa to –95 kPa are used for evaporating some VOCs with low vaporization pressures.
- Minus 95 to –99 kPa is used for VOCs with very low vaporization pressures, special vacuum pumps are required for this service and it may be desirable to reclaim the waste volatile organic compounds.

5.11.10   ADDITIONAL DESIGN ASSISTANCE REFERENCES

- Valves used in vacuum service must be suitable for use in negative pressures, ball valves are the generally accepted type; standard solenoid valves used for pressure service may not be suitable for this as most require a minimum pressure differential across the valve seat to assist in opening the valve which is supplied by the compressed gas stream and may not be available for vacuum service.

- Solenoid valves on pipelines should be selected to ensure compatibility with entrained chemicals and VOCs, these contaminants must be taken into consideration. It will be necessary to include methods for the collection and removal of any VOCs extracted by the system.

- The exhaust pipeline from any vacuum plant will be subject to a temperature increase that will be elevated as the system pressure approaches high vacuum levels. Gay Lussac's law states that the pressure of a given amount of gas held at constant volume is directly proportional to the temperature in degrees Kelvin. When calculating the physical characteristics of vacuum exhaust pipelines this is of particular importance when working with high vacuum.

KEYWORDS

- **argon**
- **carbon dioxide**
- **cryogenic liquid**
- **gaseous UHP**
- **helium**
- **instrument air**
- **nitrogen**
- **nitrous oxide**
- **oxygen**
- **vacuum**

CHAPTER 6

GAS DATA: FLAMMABLE AND TOXIC GASES

6.1 INTRODUCTION

The availability of gas specific information that is relevant to the installation of piped flammable laboratory gases is spread throughout the scientific, manufacturing, and engineering community with a little cross-reference between the various areas of expertise.

Laboratories are extremely precise in their function and operation and each has its own requirements that dictate the gases to be used, their purity and the performance required of each gas system. To supplement this there is a large volume the information available from the gas manufacturers as well as from the equipment manufacturers and each of which may vary to meet the individual company's product range. As there is no central source where this information is available to obtain at least a general basis for a laboratory gas system design we hope to provide a starting point for gas data and equipment in this chapter.

This chapter is a collection of flammable gases specific information. It has been gathered from laboratory gas manufacturers, equipment suppliers and from the manufacturers of the scientific equipment the laboratories use. Some of it comes from resolving localized issues where plant or equipment failed for unknown and obscure reasons or failed due to misinformation that had been taken from previous project designs each of which required up to date research into the uses of the gases and the associated equipment.

Common issues regularly revolved around the purity of the gases at the point of use or the provision of an incorrect gas purity being supplied after passing through the pipeline. Contamination of the gas stream by the pipeline and equipment is an all too common problem, especially where inexperienced installation companies were employed for the work.

Some of the information in the book is repetitive; this has been done purposefully. Rather than provide generic comments that are relative to a number of gases, we have attempted to provide sufficient information about a specific gas that can be obtained from one section of the chapter.

There are a number of values that are represented as approximations. Many of these are related to the contents of cylinders and the filling pressures provided by gas suppliers. These may vary between manufacturers and countries, and as such exact measurements are not possible.

All pressure ratings are provided in gauge pressure unless noted otherwise, pressure readings at absolute pressure are followed by (A).

This book is not intended to afford a 100% foolproof answer to every question; however, it proposes to provide guidelines for design engineers to locate the correct answers to questions they may have and offer suggestions about where to look for them.

6.2   ACETYLENE

6.2.1   DESCRIPTION

Acetylene is a highly flammable gas that has a strong garlic-like odor. It has a lower flammability limit of 2.5 vol. %LEL in the air and an upper limit of 81 UEL vol. % [63]. The gas has an auto-ignition temperature of 305C and may decompose explosively at pressures in excess of 101 kPa or when in contact with non-compatible materials. The gas is dissolved in Acetone in the cylinder which limits the volume of gas that can be withdrawn from the cylinder while in use. A cylinder of acetylene is able to supply a maximum flow rate that must not be exceeded; the maximum draw off must not exceed a volume greater than 10% of the volume of gas remaining in the cylinder per hour. A full cylinder of acetylene (approximately 7,000 liters) can supply 700 liters of gas for the first hour, during the second hour this volume is reduced to 630 liters of gas and reduces continuously during use. Drawing off a greater volume of gas than this will cause the Acetone to be extracted with the acetylene which in turn will contaminate the gas stream and the equipment using it and will degrade the results of the sampling procedure. As a guide it is recommended that for each Atomic Absorption Spectrophotometer (AAS) [24] a single 7,000 liter cylinder of acetylene should be provided, this can be subject to a diversity factor when multiple AASs [24] are in use that can only be provided by the user of the system who has information on the number of staff or pupil numbers using the system.

6.2.2   HAZARDS

- Highly flammable, 2.5% LEL to 81%UEL by volume in the air [63].
- Acetylene is a high-pressure gas with a cylinder pressure of 1,725 kPa at 20°C.
- Acetylene in its gaseous state is unstable at pressures exceeding 101 kPa under most circumstances and may decompose violently without notice. It must not be piped at pressures exceeding this pressure unless advised by qualified engineers with knowledge and experience working with this gas.
- Acetylene is a simple asphyxiant; to determine the levels for STEL and TWA contact should be made with National or Regional Authorities.
- Acetylene is a compressed gas that can cause physical injury to the user or other nearby individuals.
- An uncontrolled pressurized gas flow provides a source of energy that can accelerate particles and other small items to velocities that may cause physical injury to the user or other nearby individuals.
- It is highly recommended that locations using this gas have permanent gas monitoring systems with local and remote audible and visual alarms fitted and the provision of warning signs in the area where flammable, toxic, and non-respirable gases are installed.

6.2.3   LABORATORY USE

- Acetylene is the fuel gas used in Atomic Absorption Spectrophotometry (AAS) [24].

6.2.4   REFERENCES

Material Safety Data Sheets are available from the gas manufacturers that provide detailed information on this gas. The information required for any laboratory gas design must include all of the relevant properties of the gas, any compatibility issues that are applicable to the design and the acceptable pressure, flow, concentration, and other pertinent factors that are applicable to the design [19–23, 25, 26, 32, 34, 59, 60, 62].

Lubricants that are compatible with acetylene should be confirmed by the manufacturer; however, it is not recommended that these be used in parts of a pipeline that is in direct contact with the acetylene gas stream. If necessary to use lubricants the manufacturers must be contacted to confirm their products are suitable for the intended purpose and will not contaminate the purity of the acetylene. Care must be taken to ensure that any volatile lubricants or other contaminants that may enter the gas stream and contaminate the gas supply purity are specifically excluded.

6.2.5 MATERIALS OF CONSTRUCTION

6.2.5.1 COPPER TUBE

Copper tube must not be used for the construction of acetylene pipelines. Acetylene reacts with a number of materials which must not be allowed to come into contact with the gas, these include:

- Copper in its unalloyed form is unacceptable.
- Brass alloys that exceed a 65% copper content.
- Silver and its alloys that exceed a maximum of 43% silver content.
- Mercury is unacceptable at any level.

6.2.5.2 STAINLESS STEEL

- Used for the majority of acetylene pipelines for laboratory gas purity levels.
- Used for high-pressure regulators for purity levels up to Research Grade at 99.9999% or higher.

Note: Electropolished stainless steel tube is produced specifically for the electronics industry. It should be noted that this tube has the same cleanliness level as stainless steel tube cleaned for oxygen service. Electro-polishing does not improve the internal cleanliness of the pipeline. It is done specifically for the electronics industry to assist in the prevention of contamination due to outgassing from the pipeline internal surface.

6.2.5.3   ELECTROPOLISHED STAINLESS STEEL TUBE

The electropolished stainless steel tube is produced specifically for the electronics industry. It should be noted that this tube should have the same cleanliness level as certified oxygen clean tube.

The electropolished stainless steel tube is used to assist in the prevention of contamination due to outgassing where spontaneously combustible gases are piped and gases may be trapped in the pipe surface.

Note: It has been suggested that the use of this tube type will improve the flow characteristics of a pipeline. However, calculations using the Darcy Weisbach formula [21–23, 59] to calculate critical pipeline flows indicate that any increase that may occur is of little or no consequence in a laboratory gas pipeline. Any increase in flow that may be provided in using this tube is less than 0.1% of the design flow rate, the use of oxygen cleaned unpolished stainless steel tube of the same size will not impact on the working pressure or design flow. The design flow and system pressure should use a design safety margin normally set at 10% above the system working parameters.

6.2.5.4   BRASS

Brass alloys with a maximum of 65% copper content must not be used where acetylene is likely to come into direct contact.

Brass is used in the manufacture of isolation, and control valves. It is also the most common type used in the production of pressure regulators in the majority of cases for gas purity levels up to Ultra High Purity at 99.999%. The majority of laboratory bench mounted tapware is manufactured from brass with an external protective coating.

All valves used for the control of laboratory gas pipelines must be oxygen cleaned equally to the national standard for medical grade gases as a minimum; certification of cleanliness levels should be provided by the manufacturer.

6.2.5.5   SYNTHETIC TUBE

The synthetic tube has been suggested for use in high and low-pressure applications. If the synthetic tube is used for high-pressure service it must be rated for the maximum working pressure of the cylinders or pipeline to provide safe working conditions for the laboratory staff or those handling the gas cylinders. It is used in some high-pressure hoses for cylinder connections to manifolds however this tubing should be reinforced with stainless steel outer braiding or similar protection.

For low-pressure applications, the synthetic tube has a different problem including:

- The emission of volatile organic compounds used in the manufacture of the tube.
- The emission of esters used in the manufacture of the tube.
- The material used in some synthetic tubing is hygroscopic.
- Synthetic tubing and fittings may not be available in an oxygen clean state.
- Synthetic tubing may be porous to some gases, helium, and mixtures containing it are subject to this effect.

Note: The use of synthetic tube is not recommended for laboratory gas pipelines, the tubing may have materials concerns that could contaminate the gas stream. If specifically required the suppliers should provide confirmation that the material is fit for purpose.

6.2.6   PRESSURE REQUIREMENTS

For laboratory use, acetylene gas shall be piped at a maximum pipeline pressure of 101 kPa, at pressures in excess of 101 kPa the gas may become unstable. It is highly recommended that consultation with the laboratory staff is undertaken to ensure that the gas being supplied will meet with their requirements and the equipment being used. The laboratory equipment manufacturers regularly provide new or updated equipment that may require gases to be provided in a unique manner that is unable to be met by standard systems.

6.2.7   GAS SUPPLY REQUISITES

There are a number of options available for the supply of gases in laboratories. The factors that determine which type of supply source will be optimal will depend on a number of elements.

6.2.7.1   SYSTEM FLOW DEMAND

Every system will need to be designed to suit the equipment being used in the laboratory, e.g., a small single laboratory using a one bench mounted AAS [24] will require an acetylene flow that can be met by a single cylinder however an AAS [24] can be in operation for a number of hours so the supply must take that into consideration and it would be recommended that a manifold bank with a cylinder on each side would be the preferred option. A major laboratory would require considerably higher flows and it is imperative to have some information on the types and capacities of the various AAS [24]' likely to be in use.

6.2.7.2   SYSTEM PRESSURE

The pressure requirements will vary to suit the equipment being used in the laboratory however in the case of acetylene the maximum pipeline pressure is 101 kPa.

Note: System designs require that one 7,000 liter acetylene cylinder is necessary per connection to each AAS [24]. The maximum draw off rate for acetylene is approximately 10% per hour of the remaining cylinder contents. Insufficient cylinders connected to the system will allow acetone to be drawn from the cylinder and contaminate the pipeline and the AAS [24].

Acetylene cylinders should never be completely exhausted when used in laboratories, the recommended minimum residual pressure is 500 kPa for general usage, and for high flow rates, it is necessary to increase this pressure or provide a greater number of cylinders on the manifold.

6.2.7.3   SYSTEM DEPENDABILITY

Each gas system will have requirements that dictate the type of supply required; many laboratories require a permanent supply that will be online twenty for hours a day and seven days a week. This can be accomplished relatively easily for most gas supply systems using manifold cylinders. These systems need a service that will provide notification to laboratory staff that the system gas supply is in its standard operating condition or has a depleted gas supply caused by usage or an equipment failure.

To overcome this problem the cylinder manifold is provided with a backup or reserve supply that uses an automatic changeover arrangement to alternate the supply when one side of the manifold is depleted. For an acetylene cylinder supply, the number of cylinders required is duplicated with an alarm function to automatically change the supply to the reserve supply.

6.2.8   GAS SUPPLY SYSTEMS

6.2.8.1   AUTOMATIC MANIFOLDS

Automatic manifolds [30] provide a continuous supply of gas that is connected to two independent banks of gas cylinders. Each bank should have the ability to automatically provide a continuous gas supply to the laboratory to allow for a replacement of the in-use cylinder bank while operating from the reserve supply. This will need to take into consideration the time necessary to purchase the replacement cylinders and to have them delivered to the laboratory ready for use.

Purity levels specified for the gas and confirmed by the laboratory must be met by the manifold supplier; certification should be sought from the recommended supplier that the specified purity level can be provided.

Minimum cylinder pressure at changeover should be 500 kPa to prevent the acetone being drawn from the cylinder with the acetylene as the cylinder contents become depleted. This pressure will vary depending on the gas demand as long-term acetylene use for continuous periods in excess of 60 minutes may create a greater demand than the cylinder can supply (refer note in Section 6.2.7.2).

- Automatic acetylene manifolds are used in simple installations with maximum pressure requirements that do not exceed 101 kPa.
- Automatic changeover manifolds are used where continuity of supply is mandatory.

6.2.8.2   MANUAL MANIFOLDS

Manual manifolds provide a continuous supply of gas that is connected to two independent banks of gas cylinders. Each bank is manually isolated and provides a continuous gas supply to the laboratory to allow for a replacement of the in-use cylinder bank while operating from the reserve supply. This will need to take into consideration the time necessary to purchase the replacement cylinders and to have them delivered to the laboratory ready for use. The changeover from the in use to the reserve cylinders is a manual operation and requires visual inspection of the gas supply on a regular basis.

- Used where monitoring of the gas supply is not critical or can be manually operated and visually monitored. These manifolds require a manual operation to alternate the supply side of the manifold.

6.2.8.3 SINGLE CYLINDER INSTALLATIONS

Small installations that do not require large volumes of gas may also be used for systems that can be manually operated and visually monitored.

- Used when certified or gravimetric gases are required and cost can be a consideration.

6.2.8.4 BULK SUPPLY ALTERNATIVES

Cylinder packs are available for high demand systems, the number of cylinders to be used is determined by the maximum output per acetylene cylinder, exceeding this will cause acetone to be drawn from the cylinder with the acetylene, which will contaminate the gas stream and may damage the AAS [24].

When using large numbers of cylinders it is necessary to specify the pressure and maximum flow rate for the pressure regulators to meet the system design flow rate.

6.2.9 OUTLET VALVE TYPES/USES

- Laboratory taps for acetylene should be manufactured from stainless steel. Needle type valves suitably cleaned to the purity level of the gas supply are the optimal selection. Double olive compression fittings are the most common type fitted to the outlets of the taps.

Push on or serrated barb connections are unsuitable for the pressures encountered for the laboratory use of this gas.

- Valve manufacturers should be requested to confirm their range of purity levels for available types and the standard range of sizes.
- The inclusion of laboratory taps of unknown cleanliness or purity and material compatibility should be avoided at all costs.
- Medical gas outlet connections are not suitable for laboratory service.

6.2.10 PURITY LEVELS

- Instrument Grade 98.00%.

6.2.11 ADDITIONAL DESIGN ASSISTANCE REFERENCES

- Flashback arrestors should be fitted to all flammable gas pipelines; the units are available in a number of formats and sizes and must be selected to suit the application taking flows and flash arrestor types into consideration. There are models available that include a mechanical non-return valve that may create a noise as the valve opens and closes. This noise can register in the AAS [24] and cause irregularities in the printout. This type is not suitable for laboratory use.
- Adiabatic compression should be avoided wherever possible; the use of slow opening valves is recommended. Gay Lussac's law states that the pressure of a given amount of gas held at constant volume is directly proportional to the temperature in degrees Kelvin. When calculating the physical characteristics of any gas this may be of importance particularly where oxidizing, high pressure, toxic, non-respirable, and flammable gases are being installed.
- If solenoid valves are necessary the use of slow opening valve types is recommended. These can be controlled from local manually operated emergency push buttons and also interconnected to the fire trip system or any localized emergency gas monitoring system installed in the laboratory.
- Controlled pre-pressurized systems are an optional method of preventing adiabatic compression that can be caused by sudden activation of quick acting valves such as solenoid and ball valves.

- American Society for Testing Materials (ASTM) produces a number of documents that provide technical assistance for the design of gas systems. The ASTM G series publications [14–18] are of particular assistance for these types of installations.
- International and National codes of practice and regulations governing the use of dangerous goods must be referenced. In some countries, these regulations are protected by law and may not be ignored.

6.3 CARBON MONOXIDE

6.3.1 DESCRIPTION

Carbon monoxide is a highly flammable gas; it is odorless and highly toxic. It has an auto-ignition temperature of 609°C and has a lower flammability limit of 12.5 vol. %in air and an upper limit of 74 vol. % in air [63].

The gas is an asphyxiant and is extremely toxic.

6.3.2 HAZARDS

- Carbon monoxide is highly flammable, 12.5% LEL to 74% UEL by volume in the air [63].
- It is a high-pressure gas with a cylinder pressure of 20,000 kPa at 15°C.
- Carbon monoxide is highly toxic, to determine the levels for STEL and TWA contact should be made with national health authorities to determine acceptable levels and what safety protocols should be used to protect personnel who may come into contact with this gas.
- Carbon monoxide is a compressed gas that can cause physical injury to the user or other nearby individuals. An uncontrolled pressurized gas flow provides a source of energy that can accelerate particles and other small items to velocities that may cause physical injury to the user or other nearby individuals.
- Carbon monoxide is an asphyxiant that can cause nausea and vomiting followed by a loss of consciousness and death. Loss of consciousness may occur without warning in areas of oxygen concentration below 19%. Personnel may not be physically aware of the effects of the gas and may become unconscious without warning.

• It is highly recommended that locations using this gas have permanent gas monitoring systems with local and remote audible and visual alarms fitted, the provision of warning signs in the area where flammable, toxic, and non-respirable gases are installed is recommended.

6.3.3 LABORATORY USE

Carbon monoxide is used in calibration gas mixtures for the petrochemical industry, environmental emission monitoring, industrial hygiene monitors and trace impurity analyzers.

6.3.4 REFERENCES

The Material Safety Data Sheets are available from the gas manufacturers providing detailed information on this gas. The information required for any laboratory gas design must include all of the relevant properties of the gas, any compatibility issues that are applicable to the design and the acceptable pressure, flow, concentration, and other pertinent factors that are applicable to the design [19–23, 25, 26, 32, 34, 59, 60, 62].

Many lubricants are compatible with carbon monoxide; however, it is not recommended that these be used in parts of a pipeline that are in direct contact with the gas stream. If necessary to use lubricants the manufacturers must be contacted to confirm their products are suitable for the intended purpose and will not contaminate the purity of the carbon monoxide. Care must be taken to ensure that any volatile lubricants or other contaminants that may enter the gas stream and contaminate the gas supply purity are specifically excluded.

Some synthetic hose materials are not suitable for long-term exposure to carbon monoxide.

6.3.5 MATERIALS OF CONSTRUCTION

6.3.5.1 COPPER TUBE

Copper tube for use with gaseous carbon monoxide should be as supplied for medical grade gases in an oxygen clean state is suitable for the majority of carbon monoxide gas pipelines. The tube must be delivered to site pre-cleaned, dehydrated, and capped ready for use. During construction, the tube must be kept off the floor and in a location that prevents contamination from particulates or fluids that may occur on the job site. Medical Gas Grade Copper tube is used in the majority of laboratory gas pipelines.

It is recommended that hard drawn or semi-hard copper tube is used for laboratory gas pipelines unless there are incompatibility issues or it is recommended otherwise. The use of soft annealed copper pipe is not recommended for laboratory gas pipelines.

6.3.5.2 STAINLESS STEEL

There is a concern that certain grades of stainless steel are susceptible to structural stress cracking and embrittlement when used with carbon monoxide. If stainless steel is proposed for carbon monoxide using the manufacturers must be contacted for their confirmation that the material is suitable for use in this service. There are a considerable number of publications on this matter that are available for reference.

Used for special instances; however, there is no advantage in the use of Stainless steel unless there is a specific compatibility or cleanliness issue that may corrupt the gas supply. The pipeline design must not impact the gas in any way and thereby ensure the gas purity levels are unchanged when delivered at the outlet connection point.

6.3.5.3 ELECTROPOLISHED STAINLESS STEEL TUBE

The electropolished stainless steel tube is produced specifically for the electronics industry. It should be noted that this tube should have the same cleanliness level as certified oxygen clean tube.

The electropolished stainless steel tube is used to assist in the prevention of contamination due to outgassing where spontaneously combustible gases are piped and gases may be trapped in the pipe surface.

Note: It has been suggested that the use of this tube type will improve the flow characteristics of a pipeline. However, calculations using the Darcy Weisbach formula [21–23, 59] to calculate critical pipeline flows indicate that any increase that may occur is of little or no consequence in a laboratory gas pipeline. Any increase in flow that may be provided in using this tube is less than 0.1% of the design flow rate, the use of oxygen cleaned unpolished stainless steel tube of the same size will not impact on the working pressure or design flow. The design flow and system pressure should use a design safety margin normally set at 10% above the system working parameters.

6.3.5.4 BRASS

Brass is used in the manufacture of isolation and control valves. It is also the most common type used in the production of pressure regulators, in the majority of cases for gas purity levels up to ultra-high purity at 99.999%. Some laboratory bench mounted tapware is manufactured from brass with an external protective coating.

All valves used for the control of laboratory gas pipelines must be oxygen cleaned equally to the national standard for medical grade gases as a minimum; certification of cleanliness levels should be provided by the manufacturer.

6.3.5.5 SYNTHETIC TUBE

The synthetic tube has been suggested for use in high and low-pressure applications. If the synthetic tube is used for high-pressure service it must be rated for the maximum working pressure of the cylinders or pipeline and provide safe working conditions for the laboratory staff. It is used in some high-pressure hoses for cylinder connections to manifolds; however, this tubing is reinforced with stainless steel outer braiding or similar protection.

For low-pressure applications synthetic tube has a different set of problems, these include:

- The emission of volatile organic compounds used in the manufacture of the tube.
- The emission of esters used in the manufacture of the tube.
- The material used in some synthetic tubing is hygroscopic.
- Synthetic tubing and fittings may not be available in an oxygen clean state.
- Synthetic tubing may be porous to some gases, helium, and mixtures containing it are subject to this effect.

Note: The use of synthetic tube is not recommended for laboratory gas pipelines, the tubing may have materials concerns that could contaminate the gas stream. If specifically required the suppliers should provide confirmation that the material is fit for purpose.

6.3.6   PRESSURE REQUIREMENTS

For the majority of laboratory carbon monoxide gas, a pipeline pressure of 410 kPa will suffice. It is highly recommended that consultation with the laboratory staff is undertaken to ensure that the gas being supplied will meet with their requirements and the equipment being used. There are always new procedures and equipment being introduced that may require gases provided in a unique manner that is not met by standard systems.

6.3.7   GAS SUPPLY REQUISITES

There are a number of options available for the supply of gases in laboratories. The factors that determine which type of supply source will be optimal will depend on a number of elements.

6.3.7.1   SYSTEM FLOW DEMAND

Every system will need to be designed to suit the equipment being used in the laboratory, e.g., a small single laboratory may require a carbon monoxide supply which, due to the very low flow demand, may be supplied from single cylinders that can be monitored on a weekly basis. As the demand for carbon monoxide in laboratory equipment may vary considerably along with the flow demand the system design for any individual laboratory will require specific information on the pressure and flow demand of the various items of equipment being used.

6.3.7.2   SYSTEM PRESSURE

The pressure requirements will vary to suit the equipment being used in the laboratory, most manufacturers of laboratory equipment will provide a built-in pressure regulator that allows a range of supply pressures to be tolerated. This must be confirmed by the equipment manufacturer.

If there is a specific pressure required it can be provided by a local pressure regulator at the terminal outlet or at a central point in the laboratory if that is convenient, the pipeline, valves, and the pressure regulator must be sized to suit the maximum design flow required.

6.3.7.3   SYSTEM DEPENDABILITY

Each gas system will have requirements that dictate the type of supply required; many laboratories require a permanent supply that will be online twenty for hours a day and seven days a week. This can be accomplished relatively easily for most gas supply systems using manifold cylinders. These systems need a service that will provide notification to laboratory staff that the system gas supply is in its standard operating condition or has a depleted gas supply caused by usage or an equipment failure.

To overcome this problem the cylinder manifold is provided with a backup or reserve supply, in the case of cylinder supply, the number of cylinders required is duplicated with an alarm function to automatically advise staff that the change from the online supply to the reserve has occurred.

6.3.8   GAS SUPPLY SYSTEMS

6.3.8.1   AUTOMATIC MANIFOLDS

Automatic manifolds provide a continuous supply of gas that is connected to two independent banks of gas cylinders. Each bank should have the ability to automatically provide a continuous gas supply to the laboratory to allow for a replacement of the in-use cylinder bank while operating from the reserve supply. This will need to take into consideration the time necessary to purchase the replacement cylinders and to have them delivered to the laboratory ready for use.

Purity levels specified by the laboratory must be met by the manifold supplier to ensure compatibility with the gas being piped. Certification should be sought from the recommended supplier that the specified purity level can be provided.

- Automatic manifolds are used in standard installations with maximum pressure requirements that do not exceed 1,000 kPa. For higher pressures, it is recommended that the manufacturer is contacted to confirm what modification is required, is available and is compatible with the pressures specified and that the manifold will meet the expected system demand.
- Automatic changeover manifolds are used where continuity of supply is mandatory.

6.3.8.2   MANUAL MANIFOLDS

Manual manifolds provide a continuous supply of gas that is connected to two independent banks of gas cylinders. Each bank is manually isolated and provides a continuous gas supply to the laboratory to allow for a replacement of the in-use cylinder bank while operating from the reserve supply. This will need to take into consideration the time necessary to purchase the replacement cylinders and to have them delivered to the laboratory ready for use. The changeover from the in use to the reserve cylinders is a manual operation and requires visual inspection of the gas supply on a regular basis.

- Used where monitoring of the gas supply is not critical or can be manually operated and visually monitored. These manifolds require a manual operation to alternate the supply side of the manifold.

6.3.8.3   SINGLE CYLINDER INSTALLATIONS

Small installations that do not require large volumes of gas may also be used for systems that can be manually operated and visually monitored.

- Used when certified or gravimetric gases are required and cost can be a consideration.

6.3.9   OUTLET VALVE TYPES/USES

- Laboratory taps can be either diaphragm (stainless steel diaphragm) or needle type suitably cleaned to the purity level of the gas supply.

- Double olive compression fittings are the most common type fitted to the outlets of the taps. Push on or serrated barb connections are unsuitable for the pressures encountered for the laboratory use of this gas.
- Valve manufacturers should be requested to confirm their range of purity levels available for the standard range of sizes.
- Needle type valves with double olive compression fittings are common where high and low-pressure connections are required.
- It is necessary to specify that the materials of construction including all components, seals, and seats are compatible with the gases being piped.
- The inclusion of laboratory taps of unknown cleanliness or purity and material compatibility should be avoided at all costs.
- Medical gas outlet connections are not suitable for laboratory service.

6.3.10   PURITY LEVEL

- Instrument Grade 99.99%;
- High purity 99.99%;
- Ultra-High Purity 99.999%;
- Research Grade 99.9999%.

6.3.11   ADDITIONAL DESIGN ASSISTANCE REFERENCES

- Flashback arrestors should be fitted to all flammable gas pipelines; the units are available in a number of formats and sizes and must be selected to suit the application taking flows and flash arrestor types into consideration. There are some units available that include a mechanical non-return valve that may create a noise as the valve opens and closes. This noise can register in the AAS [24] and cause irregularities in the printout. This type is not suitable for laboratory use.
- Adiabatic compression should be avoided wherever possible; the use of slow opening valves is recommended. Gay Lussac's law states that the pressure of a given amount of gas held at constant volume is directly proportional to the temperature in degrees Kelvin. When calculating the physical characteristics of any gas this may

be of importance particularly where oxidizing, high pressure, toxic, non-respirable, and flammable gases are being installed.

- If solenoid valves are necessary the use of slow opening valve types is recommended. These can be controlled from local manually operated emergency push buttons and also interconnected to the fire trip system or any localized emergency gas monitoring system installed in the laboratory.
- Controlled pre-pressurized systems are an optional method of preventing adiabatic compression that can be caused by sudden activation of quick acting valves such as solenoid and ball valves.
- ASTM produces a number of documents that provide technical assistance for the design of gas systems. The ASTM G series publications [14–18] are of particular assistance for these types of installations.
- International and National codes of practice and regulations governing the use of dangerous goods must be referenced. In some countries, these regulations are protected by law and may not be ignored.

6.4 HYDROGEN

6.4.1 DESCRIPTION

Hydrogen gas is colorless, highly flammable (4 vol. % to 75 vol. % in the air [63]) and is a very light gas that diffuses rapidly. It cannot sustain life and reacts easily with other chemical substances.

6.4.2 HAZARDS

- Highly flammable, 4% to 75% by volume in the air [63].
- It is a high-pressure gas with a cylinder pressure of 13,700 kPa at 15°C.
- Hydrogen is highly toxic, to determine the levels for STEL and TWA contact should be made with national health authorities to determine acceptable levels and what safety protocols should be used to protect personnel who may come into contact with this gas.
- Hydrogen is a compressed gas that can cause physical injury to the user or other nearby individuals. An uncontrolled pressurized gas flow provides a source of energy that can accelerate particles and

other small items to velocities that may cause physical injury to the user or other nearby individuals.

- Hydrogen is an asphyxiant that can cause nausea and vomiting followed by a loss of consciousness and death. Loss of consciousness may occur without warning in areas of oxygen concentration below 19%. Personnel may not be physically aware of the effects of the gas and may become unconscious without warning.
- It is highly recommended that locations using this gas have permanent gas monitoring systems with local and remote audible and visual alarms fitted, the provision of warning signs in the area where flammable, toxic, and non-respirable gases are installed is recommended.

6.4.3   LABORATORY USE

Hydrogen is used as a fuel gas in GC [24] and in various analytical instrument applications, the most common use is as a fuel component of combustion gases for flame ionization (FID) and flame photometric detectors (FPD), spark discharge analyzers, total hydrocarbons measurement, and is also used in hydrogen gas mixtures.

6.4.4   REFERENCES

Material Safety Data Sheets are available from the gas manufacturers that provide detailed information on this gas. The information required for any laboratory gas design must include all of the relevant properties of the gas, any compatibility issues that are applicable to the design and the acceptable pressure, flow, concentration, and other pertinent factors that are applicable to the design [19–23, 25, 26, 32, 34, 59, 60, 62].

Many lubricants are compatible with hydrogen; however, it is not recommended that these be used in parts of a pipeline that are in direct contact with the gas stream. If necessary to use lubricants the manufacturers must be contacted to confirm their products are suitable for the intended purpose and will not contaminate the purity of the hydrogen. Care must be taken to ensure that any volatile lubricants or other contaminants that may enter the gas stream and contaminate the gas supply purity are specifically excluded.

6.4.5   MATERIALS OF CONSTRUCTION

6.4.5.1   COPPER TUBE

Copper tube for use with gaseous hydrogen should be as supplied for medical grade gases in an oxygen clean state is suitable for the majority of hydrogen gas pipelines. The tube must be delivered to site pre-cleaned, dehydrated, and capped ready for use. During construction, the tube must be kept off the floor and in a location that prevents contamination from particulates or fluids that may occur on the job site. Medical Gas Grade copper tube is used in the majority of laboratory gas pipelines.

It is recommended that hard drawn or semi-hard copper tube is used for laboratory gas pipelines unless there are incompatibility issues or it is recommended otherwise. The use of soft annealed copper pipe is not recommended for laboratory gas pipelines.

6.4.5.2   STAINLESS STEEL

There is a concern that certain grades of stainless steel are susceptible to structural stress cracking and embrittlement when used with hydrogen. If stainless steel is proposed for hydrogen using the manufacturers must be contacted for their confirmation that the material is suitable for use in this instance. There are a considerable number of publications on this that are available for reference. Used for some special instances however there is no advantage in the use of stainless steel, gas purity levels are unchanged.

Recommended for high-pressure service where the copper tube is not suitable due to pressure limitations taking the embrittlement issues into consideration.

6.4.5.3   ELECTROPOLISHED STAINLESS STEEL TUBE

The electropolished stainless steel tube is produced specifically for the electronics industry. It should be noted that this tube should have the same cleanliness level as certified oxygen clean tube.

The electropolished stainless steel tube is used to assist in the prevention of contamination due to outgassing where spontaneously combustible gases are piped and gases may be trapped in the pipe surface.

Note: It has been suggested that the use of this tube type will improve the flow characteristics of a pipeline. However, calculations using the Darcy Weisbach formula [21–23, 59] to calculate critical pipeline flows indicate that any increase that may occur is of little or no consequence in a laboratory gas pipeline. Any increase in flow that may be provided in using this tube is less than 0.1% of the design flow rate, the use of oxygen cleaned unpolished stainless steel tube of the same size will not impact on the working pressure or design flow. The design flow and system pressure should use a design safety margin normally set at 10% above the system working parameters.

6.4.5.4   BRASS

Brass is used in the manufacture of isolation, and control valves. It is also the most common type used in the production of pressure regulators in the majority of cases for gas purity levels up to ultra-high purity at 99.999%. Some laboratory bench mounted tapware is manufactured from brass with an external protective coating.

All valves used for the control of laboratory gas pipelines must be oxygen cleaned equally to the national standard for medical grade gases as a minimum; certification of cleanliness levels should be provided by the manufacturer.

6.4.5.5   SYNTHETIC TUBE

The synthetic tube has been suggested for use in high and low-pressure applications. If the synthetic tube is used for high-pressure service it must be rated for the maximum working pressure of the cylinders or pipeline and provide safe working conditions for the laboratory staff. It is used in some high-pressure hoses for cylinder connections to manifolds; however, this tubing is reinforced with stainless steel outer braiding or similar protection.

For low-pressure applications synthetic tube has a different set of problems, these include:

- The emission of volatile organic compounds used in the manufacture of the tube.
- The emission of esters used in the manufacture of the tube.

- The material used in some synthetic tubing is hygroscopic.
- Synthetic tubing and fittings may not be available in an oxygen clean state.
- Synthetic tubing may be porous to some gases, helium, and mixtures containing it are subject to this effect.

Note: The use of synthetic tube is not recommended for laboratory gas pipelines, the tubing may have materials concerns that could contaminate the gas stream if specifically required the suppliers should provide confirmation that the material is fit for purpose.

6.4.6 PRESSURE REQUIREMENTS

For the majority of laboratory hydrogen gas, a pipeline pressure of 410 kPa will suffice. It is highly recommended that consultation with the laboratory staff is undertaken to ensure that the gas being supplied will meet with their requirements and the equipment being used. There are always new procedures and equipment being introduced that may require gases provided in a unique manner that is not met by standard systems.

6.4.7 GAS SUPPLY REQUISITES

There are a number of options available for the supply of gases in laboratories. The factors that determine which type of supply source will be optimal will depend on a number of elements.

6.4.7.1 SYSTEM FLOW DEMAND

Every system will need to be designed to suit the equipment being used in the laboratory, e.g., a small single laboratory using one bench mounted GC [24] may require a hydrogen and nitrogen or helium supply which, due to the very low flow demand, may be supplied from single cylinders that can be monitored on a weekly basis. As the demand for this type of equipment may be in the range of 50 to 200 mL/min, a 7,000-liters cylinder of gas would provide gas for a number of weeks. A major laboratory would require higher flows; however, due to the low flow demand at any individual outlet it is imperative to have some information on the types and flow demand of the various items of equipment being used.

6.4.7.2   SYSTEM PRESSURE

The pressure requirements will vary to suit the equipment being used in the laboratory, most manufacturers of laboratory equipment will provide a built-in pressure regulator that allows a range of supply pressures to be tolerated. This must be confirmed by the equipment manufacturer.

If there is a specific pressure required it can be provided by a local pressure regulator at the terminal outlet or at a central point in the laboratory if that is convenient, the pipeline, valves, and the pressure regulator must be sized to suit the maximum design flow required.

6.4.7.3   SYSTEM DEPENDABILITY

Each gas system will have requirements that dictate the type of supply required; many laboratories require a permanent supply that will be online twenty for hours a day and seven days a week. This can be accomplished relatively easily for most gas supply systems using manifold cylinders. These systems need a service that will provide notification to laboratory staff that the system gas supply is in its standard operating condition or has a depleted gas supply caused by usage or an equipment failure.

To overcome this problem the plant or cylinder manifold is provided with a backup or reserve supply, in the case of cylinder supply, the number of cylinders required is duplicated with an alarm function to automatically advise staff that the change from the online supply to the reserve has occurred.

6.4.8   GAS SUPPLY SYSTEMS

6.4.8.1   AUTOMATIC MANIFOLDS

Automatic manifolds [30] provide a continuous supply of gas that is connected to two independent banks of gas cylinders. Each bank should have the ability to automatically provide a continuous gas supply to the laboratory to allow for a replacement of the in-use cylinder bank while operating from the reserve supply. This will need to take into consideration the time necessary to purchase the replacement cylinders and to have them delivered to the laboratory ready for use.

Purity levels specified by the laboratory must be met by the manifold supplier to ensure compatibility with the gas being piped. Certification should be sought from the recommended supplier that the specified purity level can be provided.

- Automatic manifolds are used in standard installations with maximum pressure requirements that do not exceed 1,000 kPa. For higher pressures, it is recommended that the manufacturer is contacted to confirm what modification is required, is available and is compatible with the pressures specified and that the manifold will meet the expected system demand.
- Automatic changeover manifolds are used where continuity of supply is mandatory.

6.4.8.2 MANUAL MANIFOLDS

Manual manifolds provide a continuous supply of gas that is connected to two independent banks of gas cylinders. Each bank is manually isolated and provides a continuous gas supply to the laboratory to allow for a replacement of the in-use cylinder bank while operating from the reserve supply. This will need to take into consideration the time necessary to purchase the replacement cylinders and to have them delivered to the laboratory ready for use. The changeover from the in use to the reserve cylinders is a manual operation and requires visual inspection of the gas supply on a regular basis.

- Used where monitoring of the gas supply is not critical or can be manually operated and visually monitored. These manifolds require a manual operation to alternate the supply side of the manifold.

6.4.8.3 SINGLE CYLINDER INSTALLATIONS

Small installations that do not require large volumes of gas may also be used for systems that can be manually operated and visually monitored.

- Used when certified or gravimetric gases are required and cost can be a consideration.

6.4.9 OUTLET VALVE TYPES/USES

- Laboratory taps can be either diaphragm (stainless steel diaphragm) or needle type suitably cleaned to the purity level of the gas supply.
- Double olive compression fittings are the most common type fitted to the outlets of the taps. Push on or serrated barb connections are unsuitable for the pressures encountered for the laboratory use of this gas.
- Valve manufacturers should be requested to confirm their range of purity levels available for the standard range of sizes.
- Needle type valves with double olive compression fittings are common where high and low-pressure connections are required.
- It is necessary to specify that the materials of construction including all components, seals, and seats are compatible with the gases being piped.
- The inclusion of laboratory taps of unknown cleanliness or purity and material compatibility should be avoided at all costs.
- Medical gas outlet connections are not suitable for laboratory service.

6.4.10 PURITY LEVEL

- Ultra-high purity: 99.999%.

6.4.11 ADDITIONAL DESIGN ASSISTANCE REFERENCES

- Flashback arrestors should be fitted to all flammable gas pipelines; the units are available in a number of formats and sizes and must be selected to suit the application taking flows and flash arrestor types into consideration. There are some units available that include a mechanical non-return valve that may create a noise as the valve opens and closes. This noise can register in the AAS [24] and cause irregularities in the printout, this type are not suitable for laboratory use.

- Adiabatic compression should be avoided wherever possible; the use of slow opening valves is recommended. Gay Lussac's law states that the pressure of a given amount of gas held at constant volume is directly proportional to the temperature in degrees Kelvin. When calculating the physical characteristics of any gas this may be of importance particularly where oxidizing, high pressure, toxic, non-respirable, and flammable gases are being installed.
- If solenoid valves are necessary the use of slow opening valve types is recommended. These can be controlled from local manually operated emergency push buttons and also interconnected to the fire trip system or any localized emergency gas monitoring system installed in the laboratory.
- Controlled pre-pressurized systems are an optional method of preventing adiabatic compression that can be caused by sudden activation of quick acting valves such as solenoid and ball valves.
- ASTM produces a number of documents that provide technical assistance for the design of gas systems. The ASTM G series publications [14–18] are of particular assistance for these types of installations.
- International and National codes of practice and regulations governing the use of dangerous goods must be referenced. In some countries, these regulations are protected by law and may not be ignored.

6.5 METHANE

6.5.1 DESCRIPTION

Methane is an odorless, colorless, and highly flammable gas (5 LEL vol.% to 16.4 UEL vol. % in the air [63]). It is the most common type used in the domestic gas market under the name of Natural Gas.

6.5.2 HAZARDS

- Methane is highly flammable, 5% LEL to 16.4% UEL by volume in the air [63].
- It is a high-pressure gas with a cylinder pressure of 13,700 kPa at 20°C.

- Methane is highly toxic; to determine levels for STEL and TWA contact should be made with national health authorities to determine acceptable levels and what safety protocols should be used to protect personnel who may come into contact with this gas.
- Methane is a compressed gas that can cause physical injury to the user or other nearby individuals. An uncontrolled pressurized gas flow provides a source of energy that can accelerate particles and other small items to velocities that may cause physical injury to the user or other nearby individuals.
- Methane is an asphyxiant that can cause nausea and vomiting followed by a loss of consciousness and death. Loss of consciousness may occur without warning in areas of oxygen concentration below 19%. Personnel may not be physically aware of the effects of the gas and may become unconscious without warning.
- It is highly recommended that locations using this gas have permanent gas monitoring systems with local and remote audible and visual alarms fitted, the provision of warning signs in the area where flammable, toxic, and non-respirable gases are installed is recommended.

6.5.3 USES

- Mixed with carbon dioxide, methane is used in Geiger counters and for the detector in X-ray fluorescence (XRF) as a quenching gas.
- In mixture with other hydrocarbons, methane is used as a reference point in calorimetric measurements of hydrocarbons or PCI coal.
- Methane is used in calibration gas mixtures for the petrochemical industry; environmental emission monitoring, industrial hygiene monitors and trace impurity analyzers.

6.5.4 REFERENCES

Material Safety Data Sheets are available from the gas manufacturers that provide detailed information on this gas. The information required for any laboratory gas design must include all of the relevant properties of the gas, any compatibility issues that are applicable to the design and the acceptable pressure, flow, concentration, and other pertinent factors that are applicable to the design [19–23, 25, 26, 32, 34, 59, 60, 62].

Many lubricants are compatible with methane; however, it is not recommended that these be used in parts of a pipeline that are in direct contact with the gas stream. If necessary to use lubricants the manufacturers must be contacted to confirm their products are suitable for the intended purpose and will not contaminate the purity of the methane. Care must be taken to ensure that any volatile lubricants or other contaminants that may enter the gas stream and contaminate the gas supply purity are specifically excluded.

6.5.5   MATERIALS OF CONSTRUCTION

6.5.5.1   COPPER TUBE

Copper tube for use with gaseous methane should be as supplied for medical grade gases in an oxygen clean state is suitable for the majority of methane gas pipelines. The tube must be delivered to site pre-cleaned, dehydrated, and capped ready for use. During construction, the tube must be kept off the floor and in a location that prevents contamination from particulates or fluids that may occur on the job site. Medical Gas Grade copper tube is used in the majority of laboratory gas pipelines.

It is recommended that hard drawn or semi-hard copper tube be used for laboratory gas pipelines unless there are incompatibility issues or it is recommended otherwise. The use of soft annealed copper pipe is not recommended for laboratory gas pipelines.

6.5.5.2   STAINLESS STEEL

Used for special instances; however, there is no advantage in the use of Stainless steel unless there is a specific compatibility or cleanliness issue that may corrupt the gas supply. The pipeline design must not impact the gas in any way and thereby ensure the gas purity levels are unchanged when delivered at the outlet connection point.

Stainless steel is recommended for high-pressure service where the copper tube is not suitable due to pressure limitations or availability.

6.5.5.3   ELECTROPOLISHED STAINLESS STEEL

The electropolished stainless steel tube is produced specifically for the electronics industry. It should be noted that this tube should have the same cleanliness level as certified oxygen clean tube.

The electropolished stainless steel tube is used to assist in the prevention of contamination due to outgassing where spontaneously combustible gases are piped and gases may be trapped in the pipe surface.

Note: It has been suggested that the use of this tube type will improve the flow characteristics of a pipeline. However, calculations using the Darcy Weisbach formula [21–23, 59] to calculate critical pipeline flows indicate that any increase that may occur is of little or no consequence in a laboratory gas pipeline. Any increase in flow that may be provided in using this tube is less than 0.1% of the design flow rate, the use of oxygen cleaned unpolished stainless steel tube of the same size will not impact on the working pressure or design flow. The design flow and system pressure should use a design safety margin normally set at 10% above the system working parameters.

6.5.5.4   BRASS

Brass is used in the manufacture of isolation, and control valves. It is also most commonly used in the production of pressure regulators in the majority of cases for gas purity levels up to ultra-high purity at 99.999%. Some laboratory bench mounted tapware is manufactured from brass with an external protective coating.

All valves used for the control of laboratory gas pipelines must be oxygen cleaned equally to the national standard for medical grade gases as a minimum; certification of cleanliness levels should be provided by the manufacturer.

6.5.5.5   SYNTHETIC TUBE

The synthetic tube has been suggested for use in high and low-pressure applications. If the synthetic tube is used for high-pressure service it must be rated for the maximum working pressure of the cylinders or pipeline and provide safe working conditions for the laboratory staff. It is used in some high-pressure hoses for cylinder connections to manifolds; however, this tubing is reinforced with stainless steel outer braiding or similar protection.

For low-pressure applications synthetic tube has a different set of problems, these include:

- The emission of volatile organic compounds used in the manufacture of the tube.
- The emission of esters used in the manufacture of the tube.
- The material used in some synthetic tubing is hygroscopic.
- Synthetic tubing and fittings may not be available in an oxygen clean state.
- Synthetic tubing may be porous to some gases, helium, and mixtures containing it are subject to this effect.

Note: The use of synthetic tube is not recommended for laboratory gas pipelines, the tubing may have materials concerns that could contaminate the gas stream if specifically required the suppliers should provide confirmation that the material is fit for purpose.

6.5.6   PRESSURE REQUIREMENTS

For the majority of laboratory methane gas, a pipeline pressure of 410 kPa will suffice. It is highly recommended that consultation with the laboratory staff is undertaken to ensure that the gas being supplied will meet with their requirements and the equipment being used. There are always new procedures and equipment being introduced that may require gases provided in a unique manner that is not met by standard systems.

6.5.7   GAS SUPPLY REQUISITES

There are a number of options available for the supply of gases in laboratories. The factors that determine which type of supply source will be optimal will depend on a number of elements.

6.5.7.1   SYSTEM FLOW DEMAND

Every system will need to be designed to suit the equipment being used in the laboratory, e.g., a small single laboratory using one bench mounted GC [24] may require a hydrogen and nitrogen or helium supply which, due to the very low flow demand, may be supplied from single cylinders that can be monitored on a weekly basis. As the demand for this type of equipment may be in the range of 50 to 200 mL/min, a 7,000-liters cylinder of gas would provide gas for a number of weeks. A major laboratory would require higher flows; however, due to the low flow demand at any individual outlet it is imperative to have some information on the types and flow demand of the various items of equipment being used.

6.5.7.2   SYSTEM PRESSURE

The pressure requirements will vary to suit the equipment being used in the laboratory, most manufacturers of laboratory equipment will provide a built-in pressure regulator that allows a range of supply pressures to be tolerated. This must be confirmed by the equipment manufacturer.

If there is a specific pressure required it can be provided by a local pressure regulator at the terminal outlet or at a central point in the laboratory if that is convenient, the pipeline, valves, and the pressure regulator must be sized to suit the maximum design flow required.

6.5.7.3   SYSTEM DEPENDABILITY

Each gas system will have requirements that dictate the type of supply required; many laboratories require a permanent supply that will be online twenty for hours a day and seven days a week. This can be accomplished relatively easily for most gas supply systems using manifold cylinders. These systems need a service that will provide notification to laboratory staff that the system gas supply is in its standard operating condition or has a depleted gas supply caused by usage or an equipment failure.

To overcome this problem the plant or cylinder manifold is provided with a backup or reserve supply, in the case of cylinder supply, the number of cylinders required is duplicated with an alarm function to automatically advise staff that the change from the online supply to the reserve has occurred.

6.5.8   GAS SUPPLY SYSTEMS

6.5.8.1   AUTOMATIC MANIFOLDS

Automatic manifolds [30] provide a continuous supply of gas that is connected to two independent banks of gas cylinders. Each bank should have the ability to automatically provide a continuous gas supply to the laboratory to allow for a replacement of the in-use cylinder bank while operating from the reserve supply. This will need to take into consideration the time necessary to purchase the replacement cylinders and to have them delivered to the laboratory ready for use.

Purity levels specified by the laboratory must be met by the manifold supplier to ensure compatibility with the gas being piped. Certification should be sought from the recommended supplier that the specified purity level can be provided.

- Automatic manifolds are used in standard installations with maximum pressure requirements that do not exceed 1,000 kPa. For higher pressures, it is recommended that the manufacturer is contacted to confirm what modification is required, is available and is compatible with the pressures specified and that the manifold will meet the expected system demand.
- Automatic changeover manifolds are used where continuity of supply is mandatory.

6.5.8.2   MANUAL MANIFOLDS

Manual manifolds provide a continuous supply of gas that is connected to two independent banks of gas cylinders. Each bank is manually isolated and provides a continuous gas supply to the laboratory to allow for a replacement of the in-use cylinder bank while operating from the reserve supply. This will need to take into consideration the time necessary to purchase the replacement cylinders and to have them delivered to the laboratory ready for use. The changeover from the in use to the reserve cylinders is a manual operation and requires visual inspection of the gas supply on a regular basis.

- Used where monitoring of the gas supply is not critical or can be manually operated and visually monitored. These manifolds require a manual operation to alternate the supply side of the manifold.

6.5.8.3   SINGLE CYLINDER INSTALLATIONS

Small installations, that do not require large volumes of gas, may also be used for systems that can be manually operated and visually monitored.

- Used when certified or gravimetric gases are required and cost can be a consideration.

6.5.9   OUTLET VALVE TYPES/USES

- Laboratory taps can be either diaphragm (stainless steel diaphragm) or needle type suitably cleaned to the purity level of the gas supply.
- Double olive compression fittings are the most common connection type fitted to the outlets of the taps. Push on or serrated barb connections are unsuitable for the pressures encountered for the laboratory use of this gas.
- Valve manufacturers should be requested to confirm their range of purity levels available for the standard range of sizes.
- Needle type valves with double olive compression fittings are common connections where high and low-pressure connections are required.
- It is necessary to specify that the materials of construction including all components, seals, and seats are compatible with the gases being piped.
- The inclusion of laboratory taps of unknown cleanliness or purity and material compatibility should be avoided at all costs.
- Medical gas outlet connections are not suitable for laboratory service.

6.5.10   PURITY LEVEL

- Ultra-high purity: 99.999%.

6.5.11   ADDITIONAL DESIGN ASSISTANCE REFERENCES

- Flashback arrestors should be fitted to all flammable gas pipelines; the units are available in a number of formats and sizes and must be selected to suit the application taking flows and flash arrestor types into consideration. There are some units available that include a mechanical non-return valve that may create a noise as the valve opens and closes. This noise can register in the AAS [24] and cause irregularities in the printout, this type are not suitable for laboratory use.

- Adiabatic compression should be avoided wherever possible; the use of slow opening valves is recommended. Gay Lussac's law states that the pressure of a given amount of gas held at constant volume is directly proportional to the temperature in degrees Kelvin. When calculating the physical characteristics of any gas this may be of importance particularly where oxidizing, high pressure, toxic, non-respirable, and flammable gases are being installed.
- If solenoid valves are necessary the use of slow opening valve types is recommended. These can be controlled from local manually operated emergency push buttons and also interconnected to the fire trip system or any localized emergency gas monitoring system installed in the laboratory.
- Controlled pre-pressurized systems are an optional method of preventing adiabatic compression that can be caused by sudden activation of quick acting valves such as solenoid and ball valves.
- ASTM produces a number of documents that provide technical assistance for the design of gas systems. The ASTM G series publications [14–18] are of particular assistance for these types of installations.
- International and National codes of practice and regulations governing the use of dangerous goods must be referenced. In some countries, these regulations are protected by law and may not be ignored.

6.6   METHANE (10%) IN ARGON (P10)

6.6.1   DESCRIPTION

10% methane in argon (P10) is described by different gas manufacturers as being both highly flammable and non-flammable. It is an odorless, colorless gas mixture and an asphyxiant. It is recommended to design any system using this gas as flammable and make allowances to assume that equipment included in the pipeline design is the same as would be used for hydrogen.

6.6.2   HAZARDS

- For design purposes we assume methane is highly flammable, the cylinder connection can be a left-hand thread as is used for all flammable gases.
- It is a high-pressure gas with a cylinder pressure of 13,700 kPa at 15°C.
- Methane is highly toxic; to determine the levels for STEL and TWA contact should be made with national health authorities to determine acceptable levels and what safety protocols should be used to protect personnel who may come into contact with this gas.
- 10% methane in argon is a compressed gas that can cause physical injury to the user or other nearby individuals. An uncontrolled pressurized gas flow provides a source of energy that can accelerate particles and other small items to velocities that may cause physical injury to the user or other nearby individuals.
- 10% methane in argon (P10) is an asphyxiant that can cause nausea and vomiting followed by a loss of consciousness and death. Loss of consciousness may occur without warning in areas of oxygen concentration below 19%. Personnel may not be physically aware of the effects of the gas and may become unconscious without warning.
- It is highly recommended that locations using this gas have permanent gas monitoring systems with local and remote audible and visual alarms fitted, the provision of warning signs in the area where flammable, toxic, and non-respirable gases are installed is recommended.

6.6.3   USES

- 10% methane in argon (P10) is used in electron capture equipment.

6.6.4   REFERENCES

Material Safety Data Sheets are available from the gas manufacturers that provide detailed information on this gas. The information required for any laboratory gas design must include all of the relevant properties of the gas, any compatibility issues that are applicable to the design and the acceptable pressure, flow, concentration, and other pertinent factors that are applicable to the design [19–23, 25, 26, 32, 34, 59, 60, 62].

Many lubricants are compatible with 10% methane in argon (P10); however, it is not recommended that these be used in parts of a pipeline that are in direct contact with the gas stream. If necessary to use lubricants the manufacturers must be contacted to confirm their products are suitable for the intended purpose and will not contaminate the purity of the 10% methane in argon (P10). Care must be taken to ensure that any volatile lubricants or other contaminants that may enter the gas stream and contaminate the gas supply purity are specifically excluded.

6.6.5 MATERIALS OF CONSTRUCTION

6.6.5.1 COPPER TUBE

Copper tube for use with gaseous methane mixed in Argon should be as supplied for medical grade gases in an oxygen clean state is suitable for the majority of methane argon mixture gas pipelines. The tube must be delivered to site pre-cleaned, dehydrated, and capped ready for use. During construction, the tube must be kept off the floor and in a location that prevents contamination from particulates or fluids that may occur on the job site. Medical Gas Grade copper tube is used in the majority of laboratory gas pipelines.

It is recommended that hard drawn or semi-hard copper tube be used for laboratory gas pipelines unless there are incompatibility issues or it is recommended otherwise. The use of soft annealed copper pipe is not recommended for laboratory gas pipelines.

6.6.5.2 STAINLESS STEEL

Used for special instances; however, there is no advantage in the use of stainless steel unless there is a specific compatibility or cleanliness issue that may corrupt the gas supply. The pipeline design must not impact the gas in any way and thereby ensure the gas purity levels are unchanged when delivered at the outlet connection point. Stainless steel is recommended for high-pressure service where the copper tube is not suitable due to pressure limitations or availability.

6.6.5.3   ELECTROPOLISHED STAINLESS STEEL

The electropolished stainless steel tube is produced specifically for the electronics industry. It should be noted that this tube should have the same cleanliness level as certified oxygen clean tube.

The electropolished stainless steel tube is used to assist in the prevention of contamination due to outgassing where spontaneously combustible gases are piped and gases may be trapped in the pipe surface.

Note: It has been suggested that the use of this tube type will improve the flow characteristics of a pipeline. However, calculations using the Darcy Weisbach formula [21–23, 59] to calculate critical pipeline flows indicate that any increase that may occur is of little or no consequence in a laboratory gas pipeline. Any increase in flow that may be provided in using this tube is less than 0.1% of the design flow rate, the use of oxygen cleaned unpolished stainless steel tube of the same size will not impact on the working pressure or design flow. The design flow and system pressure should use a design safety margin normally set at 10% above the system working parameters.

6.6.5.4   BRASS

Brass is used in the manufacture of isolation, and control valves. It is also the most common type used in the production of pressure regulators in the majority of cases for gas purity levels up to ultra-high purity at 99.999%. Some laboratory bench mounted tapware is manufactured from brass with an external protective coating.

All valves used for the control of laboratory gas pipelines must be oxygen cleaned equally to the national standard for medical grade gases as a minimum; certification of cleanliness levels should be provided by the manufacturer.

6.6.5.5   SYNTHETIC TUBE

The synthetic tube has been suggested for use in high and low-pressure applications. If the synthetic tube is used for high-pressure service it must be rated for the maximum working pressure of the cylinders or pipeline and provide safe working conditions for the laboratory staff. It is used in some high-pressure hoses for cylinder connections to manifolds; however, this tubing is reinforced with stainless steel outer braiding or similar protection.

For low-pressure applications synthetic tube has a different set of problems, these include:

- The emission of volatile organic compounds used in the manufacture of the tube.
- The emission of esters used in the manufacture of the tube.
- The material used in some synthetic tubing is hygroscopic.
- Synthetic tubing and fittings may not be available in an oxygen clean state.
- Synthetic tubing may be porous to some gases, helium, and mixtures containing it are subject to this effect.

Note: The use of synthetic tube is not recommended for laboratory gas pipelines, the tubing may have materials concerns that could contaminate the gas stream if specifically required the suppliers should provide confirmation that the material is fit for purpose.

6.6.6 PRESSURE REQUIREMENTS

For the majority of laboratory gas, a pipeline pressure of 410 kPa will suffice. It is highly recommended that consultation with the laboratory staff is undertaken to ensure that the gas being supplied will meet with their requirements and the equipment being used. There are always new procedures and equipment being introduced that may require gases provided in a unique manner that is not met by standard systems.

6.6.7 GAS SUPPLY REQUISITES

There are a number of options available for the supply of gases in laboratories. The factors that determine which type of supply source will be optimal will depend on a number of elements.

6.6.7.1 SYSTEM FLOW DEMAND

Every system will need to be designed to suit the equipment being used in the laboratory, e.g., a small single laboratory using one bench mounted GC [24] may require a hydrogen and nitrogen or helium supply which, due to the very low flow demand, may be supplied from single cylinders that can be monitored on a weekly basis. As the demand for this type of equipment may be in the range of 50 to 200 mL/min, a 7,000-liters cylinder of gas would provide gas for a number of weeks. A major laboratory would require higher flows; however, due to the low flow demand at any individual outlet it is imperative to have some information on the types and flow demand of the various items of equipment being used.

6.6.7.2   SYSTEM PRESSURE

The pressure requirements will vary to suit the equipment being used in the laboratory, most manufacturers of laboratory equipment will provide a built-in pressure regulator that allows a range of supply pressures to be tolerated. This must be confirmed by the equipment manufacturer.

If there is a specific pressure required it can be provided by a local pressure regulator at the terminal outlet or at a central point in the laboratory if that is convenient, the pipeline, valves, and the pressure regulator must be sized to suit the maximum design flow required.

6.6.7.3   SYSTEM DEPENDABILITY

Each gas system will have requirements that dictate the type of supply required; many laboratories require a permanent supply that will be online twenty for hours a day and seven days a week. This can be accomplished relatively easily for most gas supply systems using manifold cylinders. These systems need a service that will provide notification to laboratory staff that the system gas supply is in its standard operating condition or has a depleted gas supply caused by usage or an equipment failure.

To overcome this problem the plant or cylinder manifold is provided with a backup or reserve supply, in the case of cylinder supply, the number of cylinders required is duplicated with an alarm function to automatically advise staff that the change from the online supply to the reserve has occurred.

6.6.8   GAS SUPPLY SYSTEMS

6.6.8.1   AUTOMATIC MANIFOLDS

Automatic manifolds [30] provide a continuous supply of gas that is connected to two independent banks of gas cylinders. Each bank should have the ability to automatically provide a continuous gas supply to the laboratory to allow for a replacement of the in-use cylinder bank while operating from the reserve supply. This will need to take into consideration the time necessary to purchase the replacement cylinders and to have them delivered to the laboratory ready for use.

   Purity levels specified by the laboratory must be met by the manifold supplier to ensure compatibility with the gas being piped. Certification should be sought from the recommended supplier that the specified purity level can be provided.

- Automatic manifolds are used in standard installations with maximum pressure requirements that do not exceed 1,000 kPa. For higher pressures, it is recommended that the manufacturer is contacted to confirm what modification is required, is available and is compatible with the pressures specified and that the manifold will meet the expected system demand.
- Automatic changeover manifolds are used where continuity of supply is mandatory.

6.6.8.2   MANUAL MANIFOLDS

Manual manifolds provide a continuous supply of gas that is connected to two independent banks of gas cylinders. Each bank is manually isolated and provides a continuous gas supply to the laboratory to allow for a replacement of the in-use cylinder bank while operating from the reserve supply. This will need to take into consideration the time necessary to purchase the replacement cylinders and to have them delivered to the laboratory ready for use. The changeover from the in use to the reserve cylinders is a manual operation and requires visual inspection of the gas supply on a regular basis.

- Used where monitoring of the gas supply is not critical or can be manually operated and visually monitored. These manifolds require a manual operation to alternate the supply side of the manifold.

6.6.8.3   SINGLE CYLINDER INSTALLATIONS

Small installations that do not require large volumes of gas may also be used for systems that can be manually operated and visually monitored.

- Used when certified or gravimetric gases are required and cost can be a consideration.

6.6.9   OUTLET VALVE TYPES/USES

- Laboratory taps can be either diaphragm (stainless steel diaphragm) or needle type suitably cleaned to the purity level of the gas supply.
- Double olive compression fittings are the most common type fitted to the outlets of the taps. Push on or serrated barb connections are unsuitable for the pressures encountered for the laboratory use of this gas.
- Valve manufacturers should be requested to confirm their range of purity levels available for the standard range of sizes.
- Needle type valves with double olive compression fittings are common where high and low-pressure connections are required.
- It is necessary to specify that the materials of construction including all components, seals, and seats are compatible with the gases being piped.
- The inclusion of laboratory taps of unknown cleanliness or purity and material compatibility should be avoided at all costs.
- Medical gas outlet connections are not suitable for laboratory service.

6.6.10   PURITY LEVEL

- High purity 99.99%.

6.6.11   ADDITIONAL DESIGN ASSISTANCE REFERENCES

- Flashback arrestors should be fitted to all flammable gas pipelines; the units are available in a number of formats and sizes and must be selected to suit the application taking flows and flash arrestor types into consideration. There are some units available that include a mechanical non-return valve that may create a noise as the valve opens and closes. This noise can register in the AAS [24] and cause irregularities in the printout, this type are not suitable for laboratory use.

- Adiabatic compression should be avoided wherever possible; the use of slow opening valves is recommended. Gay Lussac's law states that the pressure of a given amount of gas held at constant volume is directly proportional to the temperature in degrees Kelvin. When calculating the physical characteristics of any gas this may be of importance particularly where oxidizing, high pressure, toxic, non-respirable, and flammable gases are being installed.

- If solenoid valves are necessary the use of slow opening valve types is recommended. These can be controlled from local manually operated emergency push buttons and also interconnected to the fire trip system or any localized emergency gas monitoring system installed in the laboratory.

- Controlled pre-pressurized systems are an optional method of preventing adiabatic compression that can be caused by sudden activation of quick acting valves such as solenoid and ball valves.

- ASTM produces a number of documents that provide technical assistance for the design of gas systems. The ASTM G series publications [14–18] are of particular assistance for these types of installations.

- International and National codes of practice and regulations governing the use of dangerous goods must be referenced. In some countries, these regulations are protected by law and may not be ignored.

KEYWORDS

- **acetylene**
- **carbon monoxide**
- **hydrogen**
- **methane**

CHAPTER 7

SAMPLE SPECIFICATION

INTRODUCTION

This sample specification excludes information relating to the electrical wiring and controls of compressor plant, vacuum plant, and the interfacing of these services with the building management systems.

There are specific gas systems such as nitrogen, argon, oxygen, liquid nitrogen, fluorine, krypton, neon, and gas mixtures referenced in this sample specification; some of which are not incorporated in this book and are shown only as examples of where specialty gases may be encountered. These rare gases are likely to be found in scientific laboratories; each specialty laboratory gas will require extensive research into the properties of these gases to ensure the pipeline design is suitable for the service and use proposed.

This sample specification is included as an example 'ONLY' for information and is not to be used or copied as a proforma for a laboratory gas system design. This sample is only to be used as an indication of the starting point for information that needs to be included in any laboratory gas specification. Further research into each of the individual gases and the technical demands of the equipment being used is required to design a safe working system.

Information that must be included should have references to National Laws, Regulations, Standards, and International Standards that must be adhered to or are used as design references for the system.

Sample Specification

SAMPLE ONLY

TENDER SPECIFICATION–LABORATORY GASES SYSTEM

Facility name and relevant details

Contents

TENDER SPECIFICATION–LABORATORY GASES SYSTEM
Project: Facility name/details

1. SCOPE OF WORKS

The project comprises the reticulation of argon, oxygen, nitrogen, ultra high purity (UHP) nitrogen, fluorine gas mix including neon, zenon, and krypton (halogen) gas and liquid nitrogen plus the supporting services to provide a safe working environment for the staff operating the various scientific equipment used in the laboratory.

1.1   ARGON SYSTEM

The argon system will be ultra high purity supplied from a single cylinder located within the laboratory. The cylinder will be connected to a high pressure point valve with flexible lead and be generally as detailed on the drawings. The cylinder will be installed in a dual cylinder safety cabinet (the nitrogen and argon will be co-located in the dual cabinet) as per the relevant clause of this specification.

The pipeline will be constructed from oxygen clean seamless stainless steel tube and connected using either orbital welding or the use of oxygen cleaned compression fittings.

The argon system pressure shall be 440 kPa and suitable for a design flow rate of up to 3.0 L/sec.

1.2   OXYGEN SYSTEM

The oxygen system will be UHP supplied from a single cylinder located within the laboratory. The cylinder will be connected to a high pressure point valve with flexible lead and be generally as detailed on the drawings. The cylinder will be installed in a single cylinder safety cabinet as per the relevant clause of this specification.

The pipeline will be constructed from oxygen clean seamless stainless steel tube and connected using either orbital welding or the use of oxygen cleaned compression fittings.

The oxygen system pressure shall be 440 kPa and suitable for a design flow rate of up to 3.0 L/sec.

1.3   UHP CYLINDER NITROGEN SYSTEM

The UHP nitrogen system will be ultra high purity supplied from a single cylinder located within the laboratory. The cylinder will be connected to a high pressure point valve with flexible lead and be generally as detailed on the drawings. The cylinder will be installed in a dual cylinder safety cabinet (the nitrogen and argon will be co-located in the dual cabinet) as per the relevant clause of this specification.

The pipeline will be constructed from oxygen clean seamless stainless steel tube and connected using either orbital welding or the use of oxygen cleaned compression fittings.

The UHP nitrogen system pressure shall be 440 kPa and suitable for a design flow rate of up to 3.0 L/sec.

1.4   NITROGEN SYSTEM

The nitrogen system will be supplied from the existing liquid nitrogen vessel connected to the gaseous pipeline downstream of the liquid/gas evaporator. The pipeline will be constructed from oxygen clean seamless stainless steel tube and connected using either orbital welding or the use of oxygen cleaned compression fittings.

The nitrogen system pressure shall be regulated in the laboratory to 440 kPa and suitable for a design flow rate of up to 3.0 L/sec.

1.5   FLUORINE MIX (HALOGEN) SYSTEM

The fluorine mix (halogen) system will be a varying mixture of <=1% fluorine, helium, argon, krypton, neon, and nitrogen or xenon supplied from a single cylinder located within the laboratory. The cylinder will be connected to a high pressure point valve with flexible lead and be generally as detailed on the drawings. The cylinder will be installed in a single cylinder safety cabinet as per the relevant clause of this specification.

The pipeline will be constructed from electropolished seamless stainless steel tube and connected by orbital welding and VCR® couplings where required.

The fluorine mix (halogen) system pressure shall be 440 kPa and suitable for a design flow rate of up to 3.0 L/sec.

Note: All stainless steel tubing must be supplied with CE Certification of oxygen cleaning stating the residual level of non-volatile hydrocarbons. CE or similar independent marking of the tubing is required. Details of the fluorine mix gas system can be found in this specification.

1.6   LIQUID NITROGEN SYSTEM

The liquid nitrogen will be supplied from the existing liquid nitrogen vessel located in the bulk liquid gas store. The pipeline will be vacuum insulated stainless steel and installed on the exterior wall of the building to enter the laboratory on level 3 as indicated on the drawing. The pipeline will terminate at a phase separator installed at high level in the MBE laboratory ready for connection to equipment by others. Details of the liquid nitrogen system can be found in this specification.

1.7   GAS SENSOR SYSTEM

The gas sensor system is to incorporate oxygen depletion and fluorine sensors in the locations indicated on the drawings. The sensors will be connected to a control module that will provide alarm functions in accordance with national guidelines.

Audible visual alarms will be provided at both sides of the entry door to each laboratory. Automatic shutdown of the argon, UHP nitrogen, piped nitrogen, liquid nitrogen, oxygen, and fluorine gas mixture will be provided.

The piped nitrogen will be isolated at a manual isolation valve and solenoid valve located at high level where the pipeline enters the laboratory. Liquid nitrogen will be isolated at the connection point where the pipeline connects to the existing bulk storage vessel. The liquid nitrogen will have a manual isolation valve and an electrically operated motorized valve located adjacent to the gas controller.

2. WORK BY OTHERS

2.1 BUILDING WORKS

The building contractor shall provide to this subcontractor the following works to enable the installation of the laboratory gas pipelines:

- Provision of gas manifold storage areas in accordance with the relevant national codes and regulations including storage security, electrical lighting and facilities for building access to allow the installation of the laboratory gas pipelines.
- Provision of amenities including 240V power boards and secure site areas for the storage of pipe and equipment necessary for the installation to proceed in accordance with the construction program.
- Provision of staff amenities including toilets and access to site.
- Provision of work lighting.

2.2 ELECTRICAL SUBCONTRACTOR

The following works shall be provided by the electrical subcontractor for this subcontractor. Provision of 240V sockets for the connections to:

- The LN2 heater at the phase separator.
- The laboratory gas alarm panel in the location nominated on the drawings.
- The gas sensor controller.

3. PIPELINE MATERIAL AND EQUIPMENT

3.1 GENERAL REQUIREMENTS

The installation shall be carried out only by experienced laboratory gas installers specifically with previous knowledge of the installation of vacuum jacketed pipework and orbital welding of Electropolished Stainless Steel. Proof of experience shall be provided prior to commencing work on site.

The following minimum requirements shall be observed during the construction of the laboratory gas systems:

- Supply all pipe to site suitably protected to prevent contamination of the internal pipe prior to installation.
- During installation all open pipe ends shall be capped when not being worked on. Recap all open end pipes when not immediately under construction and when works are stopped for extended periods.
- All equipment shall be cleaned for oxygen service, cleaning of all equipment assembled off-site shall be carried out during the manufacturing process or by the manufacturer and shall be supplied to site with certification as noted in this specification.
- Pipework shall be stored above floor level at all times.
- Use only oxygen compatible and certified Teflon tape; certification from the manufacturer shall be available on request.
- Generally all installation procedures for pipework valves and materials used in the laboratory gas installation shall be in accordance with SEMI standards, where there is no direct reference available then ISO 7396, part 1 shall apply.

3.2 PIPELINE MATERIALS

3.2.1 STAINLESS STEEL TUBE FOR ARGON, OXYGEN, AND NITROGEN

Tube shall be equal to EN 1027 D4/T3, EN 10216–5, ASTM S 213AW/269, EN 1.4401/4404 and AISI 316/316L. It shall be oxygen cleaned with a residual hydrocarbon contamination below 2.5 mg/m². CE certification from the manufacturer must be provided with the stainless steel tube and fittings.

Note: This tube is available from the manufacturer in Sweden through sales@amon.se. Extended delivery times may be expected and timely procurement should be factored into the pricing structure to ensure project completion dates.

3.2.2 STAINLESS STEEL TUBE FOR FLUORINE MIX (HALOGEN)

The stainless steel tube shall be electropolished and cleaned all in accordance with SEMI C3–0413 specification for gases 2013, SEMIE49–1103 guide for subsystem assembly and testing procedures-stainless steel systems, SEMIE49–1104 guide for high purity and ultra high purity piping performance, subassemblies, and final assemblies. Tubing shall be 316L grade stainless steel specified to ASTMA270 internally polished bore of approximately 20 Ra maximum (roughness average in micro-inches).

3.2.3   VACUUM JACKETED LIQUID NITROGEN PIPELINES

The liquid nitrogen supply system shall be a vacuum insulated pipeline specially designed for use with liquid nitrogen working at a design pressure of no less than 1500 kPa. The internal line is to be 20 mm OD tube with 75 mm OD tube on the outer. The vacuum pipeline is to be either 304 or 316 grade stainless steel. The outer is to have a polished finish, nominally 300 grit, to assist in reflecting heat from the pipeline.

Each section of line will have its own integral annular vacuum. The vacuum pressure must be less than 10×10^{-6} mmHg (10 micron), at time of installation. The inner tube is to be wrapped with no less than 32 layers of radiant heat insulating material, nominally aluminized Mylar or multi-layer insulation. Each section or "spool" is to also include a sachet of molecular sieve to act as a moisture getter material.

Each section, change of direction or connection will be joined with a bayonet connection to prevent heat in leak, and simplify installation.

Each section must be helium mass spectrometry tested to less than 1 $\times 10^{-8}$ atmosphere cc/sec, to ensure a high vacuum level, and longevity of the vacuum. Vacuum integrity must be maintained. The vacuum must be guaranteed to be less than 20 micron after 12 months or the duration of the defects liability period.

The supplier of the vacuum jacketed pipeline must determine the appropriate locations of elbows and bayonet connections. Allowance for expansion and contraction of the inner line must be incorporated with the inclusion of internal expansion bellows. These must be able to take up the contraction of the line between ambient and minus 200°C temperatures.

External stainless steel tube shall be designed with expansion bellows or similar to allow movement of the outer pipeline to take atmospheric ambient conditions into consideration.

Penetrations between external and internal walls by pipework are to be covered with stainless cover plates. Pipe supports are to be stainless steel tube clamps, if other metals are used plastic or rubber strips must be used to prevent metal-to-metal contact.

3.2.4   NON-VACUUM INSULATED SECTIONS

Where non-vacuum insulated sections of pipework exist, the parts exposed to the atmosphere shall be insulated with Polyurethane, Armaflex or equivalent of suitable thickness to prevent the accumulation of ice. Examples of these connections are:

- Connection to the liquid nitrogen header including repairs to the existing header during installation.
- Vent line connection downstream of the exhaust heater and the vent pipe.
- Uninsulated connections to the phase separator, if applicable.

Note: These connection points shall be insulated then wrapped and sealed with waterproof tape to prevent the formation of condensation.

3.2.5   VCR® COUPLINGS, VALVE FITTINGS AND CONSTRUCTION

Connections using VCR® Couplings and Swagelok or equivalent compression fittings shall be provided where applicable and shall be supplied in the same oxygen cleaned condition as the Stainless Steel tube. Certification of cleanliness shall be provided as for pipeline materials. Cleanliness levels shall be equal to Swagelok ultra high purity process specification SC–01 (Specification SCS–00001 Rec C).

3.2.6   IDENTIFICATION OF PIPELINES

Identifying labels shall be fixed to laboratory gas pipelines at all junctions, terminations, pressure regulation and measuring points, at the entry and exit of bulkheads and wall penetrations and at intervals of 3 meters throughout their length. Identification labels shall be fixed to the pipeline and shall be in accordance with AS 2896–2011 Clause 3.6.1.2 or Clause 3.6.1.3.

Where pipelines are inaccessible or where the pipeline is not normally accessible or where access to the pipeline is restricted (e.g., in ceilings or service ducts) it shall be identified by a band of not less than 70 mm in width, colored as specified for the valve end of laboratory gas cylinders in AS 1944 and the name of the particular gas distributed by the pipeline qualified by the word 'Laboratory' as specified in AS 2896–2011, Figure 3.3 and Labels shall be as noted in AS 2896–2011, Table 3.1.

3.3   VALVES AND REGULATORS

3.3.1   HIGH PRESSURE POINT VALVES

Brass high pressure point valves and flexible leads shall be provided for the argon, oxygen, and nitrogen cylinder connections and shall be gas specific where possible and compatible with the gases being installed. Regulator connections shall be as per CGA configuration or as per National Standards where required.

Stainless steel high pressure point valve and flexible lead shall be provided for the fluorine mix gas and shall be gas specific. Regulator connections shall be CGA configuration as required. The fluorine mix gas high pressure point valve shall be fluorine passivized at the time of manufacture.

3.3.2   ULTRA HIGH PURITY REGULATORS

Ultra High Purity stainless steel regulators shall be provided for all gases. Regulators shall be dual stage with diaphragms and all whetted parts manufactured from 316 stainless steel. The regulators shall be suitable for inlet pressures up to 21,000 kPa with an operating temperature range of $-40°C$ to $60°C$ and argon integrity of 1×10^{-9} cc/sec.

The regulator shall be fitted with a safety relief valve suitable for connection of an exhaust pipe to be vented to a location outside the building.

The regulator shall be fitted with a purge valve and suitable for connection of an exhaust pipe to be vented to a location outside the building.

The fluorine mix gas regulator shall be fluorine passivized at the time of manufacture.

Note: Extended delivery times may be expected and timely procurement should be factored into the pricing structure to ensure project completion dates.

3.3.3   TERMINAL OUTLET VALVES

Termination isolation valves shall be provided in the locations nominated on the drawings as per the following chart (Table 7.1).

TABLE 7.1   Laboratory Gas Terminal Schedule

Gas	Manufacturer	Size	Type	Material
Nitrogen	Swagelok or equal	6 mm	¼ turn ball valve	Stainless Steel
Argon	Swagelok or equal	6 mm	¼ turn ball valve	Stainless Steel
Oxygen	Swagelok or equal	6 mm	¼ turn ball valve	Stainless Steel
Fluorine Mix	Swagelok or equal	6 mm	¼ turn ball valve	Stainless Steel
Liquid Nitrogen	Four (4) blanked connection points		Phase separator	Stainless Steel

Note 1: The fluorine mix isolation valve shall be passivized at time of manufacture.

Note 2: Liquid nitrogen isolation valves where installed shall be vacuum insulated or shall be insulated with Polyurethane, Armaflex or equivalent of suitable thickness to prevent the accumulation of ice.

Note 3: Labels shall be provided at each terminal location indicating the gas name and pressure, similar gases shall be identified separately, e.g., Piped nitrogen and UHP nitrogen.

3.4   GAS CYLINDER SAFETY CABINETS

Supply and install 2 off 1-cylinder cabinets and 1 off 2-cylinder cabinet. Cabinets shall be of all welded one piece construction in 11-gauge (3 mm) steel with epoxy painted textured finish on the outside and smooth on the inside.

The cabinets shall be equal to the Duperthal Lab systems Supreme Line Type G90 Storage Cabinets.

Duperthal's Supreme line of safety storage cabinets used for the storage, use, and emptying of pressurized gas cylinders. The Supreme line incorporates a fire resistance of 90 minutes for temperature increase of 50 Kelvin measured at the neck of the gas cylinder. Interior height shall enable ease of installation of gas cylinders and provide OHS compatible access to the cabinet and operation of the gas cylinder fittings. The standard integrated installation rails shall be adjustable in height for convenience. Pressurized gas cylinders are to be secured against accidentally falling over by means of standard integrated cylinder holder with retaining belts. Pipes and cables can be directly laid from the gas fitting through the cabinet ceiling to the outside.

A fire resistance of 90 minutes is required: the model selected should prevent the spread of a fire within the laboratory and potentially provide protection for a minimum period of 90 minutes to allow staff to be evacuated and fire-fighters to attend to the hazard.

The cabinet shall be constructed from high-quality powder-coated sheet steel equal to the Duperthal Type G90 storage cabinets and be constructed with smooth cabinet surfaces with no protruding hinges or covers.

Cabinets should have:

- Fire resistance of 90 minutes for temperature increase of 50 Kelvin;
- Wing doors sheet steel, powder coated;
- Outer carcass sheet steel, powder coated;
- Inner carcass made from high-quality décor panels;
- Two installation rails for holding gas fittings;
- One cylinder holder each with one retaining belt per standing space;
- Integrated drilling template;
- Standard integrated installation rails adjustable in height;
- Ventilation and extraction in the whole cabinet due to slits in the air supply and exhaust air ducts;
- Door opening angle continuously adjustable up to 170°C;
- Exceeds requirements of Australian Standard AS/NZS 1940;
- Type tested and classified to European standards DIN EN 14470–1 & DIN EN 14727.

3.5   GENERAL INSTALLATION REQUIREMENTS FOR PIPELINES

3.5.1   MAINTAINING CLEANLINESS DURING CONSTRUCTION

It is of the utmost importance that the integrity of the internal surfaces of the pipeline, valves, and fittings is maintained. To facilitate this these pipelines, valves, and fittings shall be kept in protective wrapping until such time as the installation is due to proceed. During construction all open ends of pipe, valves, and fittings shall be kept sealed unless being worked on.

Any pipe, valves or fittings left open to contamination shall be removed from site and replaced with new sealed parts. Contaminated parts shall not be reused for the project.

3.5.2   ORBITAL WELDING PROCEDURES AND SPECIFICATIONS

The weld joint should be purged with a gas to eliminate or minimize the appearance of oxidation on the inner surface. Cracks and crevices, porosity, welding dross (slag), which would interfere with complete fusion of the joint, should all be avoided. The weld bead should be of uniform thickness, without thin spots and the arc should not differ from side to side to such a degree that could result in a lack-of-fusion defect.

Fusion butt welds shall be completely penetrated around the entire circumference, without excessive concavity or convexity of the outer or inner surface.

Alignment of the weld components should be such that no ridge is formed on the inner surface of the weld joint. For proper alignment, the dimensional tolerances of the tubing and fitting ends must meet material specifications for diameter, wall thickness variations and ovality. In addition, a good machined square-end preparation is a prerequisite for consistent high-quality orbital welds.

3.5.3   LOCATION AND INSTALLATION

All pipework shall be installed at a nominal height of 2,700 mm above FFL or as indicated on the drawings. This shall be maintained throughout all horizontal pipe runs. The height of this pipework will be governed by the installation of the vacuum jacketed pipework and this in turn shall be dictated by the location and installation of the liquid nitrogen phase separator which must be installed at the highest point of the horizontal liquid nitrogen pipeline in the MBE laboratory. The selected zone for the installation of the liquid nitrogen and gaseous systems in the laboratory is 2,700 mm to the underside of the pipeline above FFL.

The pipeline design shall ensure that there are no high points or locations that facilitate the accumulation of gas pockets.

3.5.4 PIPELINE SUPPORTS

Laboratory gas pipelines shall be supported at intervals sufficient to prevent sagging or distortion in accordance with Table 7.2. Supports shall be of the correct strength to prevent the pipeline being moved accidentally from its position. Supports shall be constructed of metal and made so that they are corrosion resistant. For this purpose, they may be treated or sleeved. The support shall not cause corrosion of the pipeline and protective insulation shall be supplied for all services where dissimilar metals may be encountered.

A common support bracket of sufficient strength is permitted to independently support each gas pipeline. U-bolt pipe supports shall not be used. Where vertical pipes are exposed in rooms they shall be secured at floor and ceiling; pipes up to 25 mm shall have at least two intermediate supports.

TABLE 7.2   Pipeline Bracket Support Spacing

Nominal pipe size OD	Maximum horizontal space (m)	Maximum vertical space (m)
15	1.5	1.5
20	1.5	2
25	2	2.5

Supports for the liquid nitrogen shall be as recommended by the manufacturer.

4.   GAS SENSING SYSTEM

The gas sensing system shall include a central controller to receive inputs from the various sensors and inputs throughout the laboratory and provide alarms and controls for local equipment as well as clean terminals for connection to the BMS.

4.1   GAS CONTROLLERS

The Gas Controller shall be capable of continuous monitoring and warning of toxic, atmospheric oxygen depletion and enrichment.

It shall include a minimum of 12 analogue gas inputs using 4 to 20 mA signals and digital inputs from the emergency push buttons in each laboratory. The analogue inputs shall provide three alarm set points per input.

Output relays with clean NO/NC contacts shall be provided for control of:

- Motorized isolation valve for the liquid nitrogen supply at the connection to the liquid nitrogen header adjacent to the LN2 vessel.
- Solenoid valves for the isolation of the UHP nitrogen, piped nitrogen, argon, oxygen, and fluorine mix gases located in the gas cabinets.

Alarm outputs shall be provided for:

- BMS connections (NO/NC clean terminals for connection by the BMS contractor).
- Operation of the solenoid and motorized valves on each gas system.
- Independent audio visual alarms for each laboratory.

The gas controller shall include a self-checking system to monitor the integrity of the analogue inputs to detect short-circuit and open circuit. The gas controller is to be fitted with a 24V DC power supply plus backup to operate the solenoid and motorized valves and all devices incorporated in the Laboratory Gas system.

It shall provide sufficient output connections to operate the audible visual alarms for oxygen depletion/enrichment and fluorine levels, located either side of all entry doors to the laboratories.

4.2 GAS SENSORS

The gas sensors should be supplied from the same manufacturer as the gas controller without puts using 4 to 20 mA signals, where specific sensors are provided they shall be gas specific and pre-calibrated prior to delivery to site. The numbers of sensors in each location shall be determined by the dimensions of the laboratory with the sensors located as indicated on the drawings or in locations recommended by the manufacturer. The numbers of sensors indicated on the drawings are indicative only and the final location and number of sensors shall be as recommended by the manufacturer to suit the type being offered.

4.2.1   OXYGEN DEPLETION AND ENRICHMENT ALARM

The oxygen depletion and enrichment alarm shall combine an oxygen sensor and transmitter including digital measurement value processing and temperature compensation for the continuous monitoring of the oxygen concentration in the ambient air. The calibration routine with selective access release is to be integrated in the transmitter. It shall have a standard analogue 4–20 mA or 2–10 V DC using an RS–485 interface. Two relays with adjustable switching thresholds are required. The unit shall have integrated controls including:

- Dual level oxygen depletion alarm (for monitoring the laboratory gas atmosphere and the gas cabinet internal environments).
- Oxygen enrichment alarm (for monitoring the gas cabinet internal environment).
- Automatic synchronization on multi-sounder system.
- Continuously rated.
- Stainless steel fixings.
- Mounting via internal fixing positions or via external mounting lugs.
- Duplicate cable terminations (in and out for daisy-chain installations).
- Available with custom tone configurations and frequencies.

4.2.2   FLUORINE ALARM

The fluorine transmitter shall be capable of providing digital measurement value processing and temperature compensation for the continuous monitoring of the ambient air to detect fluorine concentrations. Integrated in the transmitter there shall be a calibration routine with selective access release. The unit shall be equipped with standard analogue output 4–20 mA or 2–10 V DC and an RS–485 interface. Two relays with adjustable switching thresholds shall be provided.

The unit shall be designed for the detection of fluorine within a wide range of industrial and commercial applications including:

- Digital processing of the measurement values;
- Temperature compensation;
- Continuous monitoring;
- Low zero point drift;

- Modular plug-in design;
- Reverse polarity protected, overload, and short-circuit proof;
- Serial interface RS–485;
- IP65 protection;
- Manual calibration via potentiometer; and
- Manual addressing for RS–485 mode.

4.3 AUDIBLE/VISUAL ALARMS

Audible/visual alarms shall be provided in the locations indicated on the drawings, these shall be installed at the entry and exit doorway to each local area being served and in any isolated rooms within the laboratory. The unit shall incorporate a variable volume alarm sounder and beacon with the following inclusions:

- Automatic synchronization for a multi-sounder system;
- Continuously rated;
- Stainless steel fixings;
- Mounting via internal BESA compatible fixing positions or via external mounting lugs;
- Duplicate cable terminations (in and out for daisy-chain installations);
- Custom tones and frequencies.

4.4 EMERGENCY CONTROLS

Emergency push buttons will be provided by the electrical trade subcontractor. The push buttons will have independent clean contacts provided for connection to the gas sensor controller.

4.5 CONTROLS AND WIRING FOR AUTOMATIC AND EMERGENCY SHUTOFF SYSTEMS

4.5.1 INDEPENDENT LIQUID NITROGEN ISOLATION AND CONTROL

An independent isolation switch shall be provided adjacent to the gas controller to manually operate the liquid nitrogen supply control valve.

The liquid nitrogen control will be a motorized stainless steel valve suitable for liquid nitrogen service located at the connection point to the existing liquid nitrogen connection adjacent to the vacuum insulated vessel.

A manual isolation valve shall also be fitted prior to the motorized valve for service and maintenance should it be necessary.

Relief valves with piped exhausts shall be fitted to the liquid nitrogen pipeline at any location where sections of pipeline may cause isolation of the liquid nitrogen.

4.5.2   LOCAL ISOLATION AND CONTROL FOR NITROGEN, ARGON, OXYGEN, AND FLUORINE MIX GAS

Solenoid valves will be provided within the gas control cabinets. The solenoid valves will be:

- Stainless steel construction;
- Oxygen cleaned for the service complete with certification from the manufacturer;
- Normally closed;
- 24VDC continuously rated; and
- 12 mm line size.

Note: Solenoid valves will be passivized at time of manufacture for fluorine mix only.

4.6   WIRING INSTALLATION

The works included shall provide electrical services for the operation of all controls, plant, and equipment in accordance with national wiring regulations and the manufacturer's specifications.

All interconnecting wiring between the gas controller and gas sensors, emergency push buttons, audible/visual alarms and controls is to be included in the tender price.

110/240V supply will be provided by others to a single location adjacent to the gas controller.

5. TESTING AND COMMISSIONING

In this specification the requirements for flow rate and the pressure for the gas delivered are expressed at NTP (Normal Temperature and Pressure).

Except for those tests in which the gas is specified, purging, and testing shall be carried out using dry nitrogen only, filling the system with the specific gas shall not be carried out until all testing and commissioning procedures are complete and the system is ready for use.

Before any testing is carried out every terminal unit in the system under test shall be labeled to indicate that the system is under test and the terminal unit must not be used.

The resolution and the accuracy of all measuring devices used for testing shall be appropriate for the values to be measured.

All testing and measuring devices used for certification shall be approved by a nationally certified laboratory or equivalent and shall be calibrated at appropriate intervals. Copies of certification shall be provided for all testing and commissioning equipment where appropriate.

When the results of a test do not meet the acceptance criteria, remedial work shall be carried out and the tests repeated as necessary.

5.1   TESTS, CHECKS, AND PROCEDURES BEFORE USE OF THE SYSTEM

The following tests and procedures shall be carried out:

- Pressure test for leakage;
- Test for cross connection;
- Test for gas flow;
- Test for gas pressure;
- Test for particulate contamination;
- Test for mechanical integrity; and
- Commissioning.

These tests shall be witnessed by the representative of the facility, the design engineer and any regulatory authority having controlling approvals and shall be carried out prior to completion and before handover of the system. Copies of all witnessed documentation shall be provided to those personnel present during the testing procedures and a permanent copy shall form part of the Operational and Maintenance Manual for the project.

5.2 SYSTEM PRESSURE TESTS

After installation of terminal outlets and before any parts are concealed below benches or in walls, each section of the piping system shall be subjected to a pressure test by the installer in accordance with the following:

a) Argon, piped nitrogen, UHP nitrogen and oxygen shall be tested at 1.5 times the working pressure of the system, the only allowable pressure variation in a 4 hour period shall be that caused by variations in the ambient temperature adjacent to the pipeline system.

b) Liquid nitrogen shall be tested at the maximum manufacturer's test pressure or 1,750 kPa and the only allowable pressure change in a 4 hour period shall be that caused by variations in the ambient temperature adjacent to the pipeline system.

Each service shall be purged with dry nitrogen and filled to the test pressure, this pressure shall be held for a period of five minutes to allow for any variation due to adiabatic compression that may have occurred and then adjusted as necessary to the predetermined test pressure.

The systems tests shall be witnessed by the authorized representative of the facility and the technician carrying out the pressure test on the relative documentation indicating the date and time the pressure test was applied after which the test gas shall be isolated and connecting pipework removed from the pipeline for a minimum period of 4 hours.

Following the expiry of the predetermined time the pressure test equipment shall be inspected; there should be no variation in the test pressure with the exception of any variation that may be due to alteration in the ambient temperature or conditions.

Should there be any variation in the pressure that cannot be attributed to those noted above, the system shall be checked by the installer and after rectification has been undertaken the system shall be retested.

During these tests any equipment such as switches, pressure gauges, regulators or other equipment that may be damaged or are unsuitable for the test pressure shall be removed or isolated from the system.

During these tests liquid nitrogen supply plants and cylinder supply manifolds shall not be connected to the pipelines under test.

Note: Hydraulic testing shall not be used.

5.3   CROSS CONNECTION TEST

Testing for cross-connections in the piping systems installed shall include all pipelines and valves from the gas supply source to terminal units and shall be in accordance with the following:

a)   All systems shall be depressurized with the exception of the one under test which shall be filled with the test gas to the working pressure of the system.

b)   Each terminal unit of every gas piping system shall be inspected to determine that the test gas is being dispensed only from the terminal units of the system under test. After completion of each test the tested system shall be depressurized prior to moving to the next gas system.

c)   Repeat these steps for each gas in the system. The results shall be recorded and witnessed and the test sheet shall form part of the system manual on the completion of the project.

5.4   FLOW, PRESSURE, AND PARTICULATE TEST

All gas systems shall be filled with dry nitrogen to the design working pressure of the respective system. Every terminal outlet shall be tested for pressure and flow to verify that the performance is in accordance with this specification.

These tests shall be carried out using dry nitrogen as the test gas and as follows:

a)   Apply the test pressure gauge to the terminal unit and record the static pressure.

b)   Attach the flow meter to the terminal unit and confirm that the specified flow rate can be provided; the terminal unit may not necessarily be fully open during this test.

c)   Using a particle counter at the terminal unit in accordance with the manufacturer's instructions set the flow rate at 1.5 times the specified flow rate for a period of 30 seconds. The particulate count should be less than that specified in ISO 8573–2010 Part 1 Compressed Air Purity Classes for Clause 5.2 Particulate Classes Table 1, Class 1. The test methods used should be as set out in ISO 8573–2001-Part 4 Test Methods for Solid Particle Content.

d) Repeat the test for each terminal outlet in the system. The results shall be recorded and witnessed and the test sheet shall form part of the system manual on the completion of the project.

5.5   MECHANICAL INTEGRITY TESTING

Prior to handover all equipment provided as part of this project shall be physically operated in the presence of an authorized representative of the University.

Test sheets shall be provided for all tests carried out. These shall be witnessed by the representative of the facility and the contractor/subcontractor. The following tests are to be carried out:

- All mechanical and electrically operated valves shall be opened and closed.
- All pressure switches operated and alarm functions confirmed.
- All gas sensors shall be tested and the set limits confirmed.
- The gas sensor controller shall be tested and all displayed results confirmed.

5.6   COMMISSIONING

Each system shall be purged with its own designated gas for a sufficient time to fill the system with that gas and to reach full working pressure. Attention shall be paid to purging each outlet to ensure no pockets of test gas remain, beginning with the outlet closest to the source and ending at the most remote outlet. During this procedure the vented gases should be exhausted to a secure location outside the building.

6.   OPERATING AND MAINTENANCE MANUAL

Operating and maintenance manuals will be provided in triplicate and on DVD/CD and original documents only will be used in the manuals. One copy will be provided for approval and shall be submitted at least 6 weeks before the programmed completion date of the project. Practical completion will not be awarded until approved manuals have been submitted.

6.1 GENERAL DESCRIPTION

The operating and maintenance manual will include headings as follows:

- Laboratory oxygen system;
- Piped nitrogen system;
- UHP nitrogen system;
- UHP argon system;
- Fluorine gas mixture; and
- Liquid nitrogen system.

6.2 EQUIPMENT DETAILS

This section will include sub clauses with brief descriptions of the equipment included in that system giving manufacturers details, model numbers and types that have been included. Where equipment is used for multiple systems (e.g., gas sensors) cross references shall be included to indicate the common usage.

Each heading shall include manufacturers operating and maintenance details for the equipment provided as part of the sub contract which will allow the maintenance staff the necessary information to make minor adjustments to the equipment should they be required.

This section shall have sub paragraphs as follows:

- Gas supply manifolds;
- Valves and controls;
- Liquid nitrogen system; and
- Gas sensor system and controls.

6.3 SYSTEM DESCRIPTION

A detailed explanation of each laboratory gas system shall be provided for the following systems giving explanations of the operation of the gas system for daily use:

- Laboratory oxygen system;
- Piped nitrogen system;
- UHP nitrogen system;
- UHP argon system;
- Liquid nitrogen system;
- Fluorine gas mixture;
- Gas sensor controller and sensor; and
- Safety measures and controls for each gas system.

6.4 TESTING AND COMMISSIONING RESULTS

Copies of all tests carried out during construction shall be included in this section of the manual including:

- System pressure test;
- Cross connection test;
- Flow and pressure test;
- Mechanical integrity test; and
- Settings of all controls, operating pressures, safety relief valve settings.

6.5 DRAWING REGISTER

- Complete list of all construction drawings and diagrams.
- Complete set of drawings in A3 format.
- DVD or CD folder with all drawings in DWG and PDF format and manually written to CD.

7. TENDER SCHEDULES

7.1 PRICE BREAKUP

The price breakup will be determined by the design engineer; pricing may be on a per gas basis or may be based around the equipment or on a milestone structure (Table 7.3).

Table 7.3. Pricing Schedule

Item	Schedule rate
Liquid nitrogen pipeline	$
Piped nitrogen	$
UHP nitrogen	$
UHP argon	$
UHP oxygen	$
Fluorine mix	$
Gas cabinets	$
Gas sensor system and wiring	$
Gas isolation system and wiring	$
Drawings	$
Total =	$

7.2 EQUIPMENT SUPPLIERS

The supplier listing may also require plant and equipment capacities, delivery timetable, dimensions all of which may be information that will ensure the proposed plant will be suitable for installation in the areas proposed by the architect (Table 7.4).

TABLE 7.4   Equipment Manufacturers Schedule

Item	Manufacturer/Make	Model
Liquid nitrogen pipeline		
Stainless steel tube supplier		
Stainless steel fitting supplier		
Gas cabinet supplier		
Gas manifold supplier		
Gas terminal outlet supplier		
Oxygen sensor supplier		
Fluorine sensor supplier		
Audible/visual alarm supplier		
Gas controller supplier		
Onsite installation contractor		
Solenoid supplier		
Isolation valve supplier		

8. DESIGN STANDARDS

The design of the gas monitoring system shall be in accordance with the following international and national standards:

- MSDS for piped gases.
- ASTM 632–04 Standard Specification for Seamless and Welded Austenitic Stainless Steel Tubing for General Service.
- ASTM A270/A270M–10 Standard Specification for Seamless and Welded Austenitic Stainless Steel Sanitary Tubing.
- ASTM Flammability and Sensitivity in Oxygen Enriched Atmospheres.
- ASTM G 127–95 Standard Guide for the Selection of Cleaning Agents for Oxygen systems.
- ASTM G 63–99 Standard Guide for Evaluating Nonmetallic Materials for Oxygen Service.
- ASTM G 88–05 Standard Guide for Designing systems for Oxygen Service.
- ASTM G 93–3 Standard Practice for Cleaning Methods and Cleanliness Levels for Material and Equipment Used in Oxygen-Enriched Environments.
- ASTM G 94–05 Standard Guide for Evaluating Metals for Oxygen Service.
- ASTM Technical Training Reference Books.
- IEC 61508 2010–04 Part 1: Functional safety of electronic/electronic/programmable electronic safety-related systems, General requirements. Geneva, Switzerland: International Electrotechnical Commission; 2010.
- IEC 61508 2010–04 Part 2: Functional safety of electronic/electronic/programmable electronic safety-related systems, Requirements for electrical/electronic/programmable electronic safety-related systems. Geneva, Switzerland: International Electrotechnical Commission; 2010.
- IEC 61508 2010–04 Part 3: Functional safety of electronic/electronic/programmable electronic safety-related systems, Software requirements. Geneva, Switzerland: International Electrotechnical Commission; 2010.
- IEC 61508 2010–04 Part 4: Functional safety of electronic/electronic/programmable electronic safety-related systems, Definitions,

and abbreviations. Geneva, Switzerland: International Electrotechnical Commission; 2010.

- IEC 61508 2010–04 Part 5: Functional safety of electronic/electronic/programmable electronic safety-related systems, Examples of methods for the determination of safety integrity levels. Geneva, Switzerland: International Electrotechnical Commission; 2010.

- IEC 61508 2010–04 Part 6: Functional safety of electronic/electronic/programmable electronic safety-related systems, Guidelines on the application of IEC 61508–2 and IEC 61508–3. Geneva, Switzerland: International Electrotechnical Commission; 2010.

- IEC 61508 2010–04 Part 7: Functional safety of electronic/electronic/programmable electronic safety-related systems, Overview of techniques and measures. Geneva, Switzerland: International Electrotechnical Commission; 2010.

- IEC 61511–1: 2016–1 Functional safety–Safety instrumented systems for the process industry sector–Part 1: Framework, definitions, system, hardware, and application programming requirements. 2nd ed. Geneva 20, Switzerland: International Electrotechnical Commission; 2016.

- IEC 61511–2016–2 Functional safety–Safety instrumented systems for the process industry sector. Part 2: Guidelines in the application of IEC 61511–1. 1st ed. Geneva 20, Switzerland: International Electrotechnical Commission.; 2003.

- IEC 61511–2016–1 Functional safety–Safety instrumented systems for the process industry sector–Part 3: Guidance for the determination of the required safety integrity levels. 2nd ed. Geneva 20, Switzerland: International Electrotechnical Commission; 2017.

- IEC 62061: 2005 Safety of machinery–Functional safety of safety-related electrical, electronic, and programmable electronic control systems. 1st ed. Geneva: International Electrotechnical Committee; 2015.

- ISA TR84.00.02: 2015 Safety Integrity Level (SIL) Verification Of Safety Instrumented Functions. Research Triangle Park, N.C.: ISA-International Society of Automation; 2015.

- ISA TR84.00.02–3: 2002 Safety Instrumented Functions (SIF)–Safety Integrity Level (SIL) Evaluation Techniques–Part 3: Determining the SIL of A Sif via Fault Tree Analysis. International Society of Automation; 2002.

- ISO 13109 Pipeline Materials.
- SEMI C3–0413 Specification for Gases 2013.
- SEMI E49–1103 Guide for Subsystem Assembly and Testing Procedures-Stainless Steel systems.
- SEMI E49–1104 Guide for High Purity and Ultra High Purity Piping Performance, Subassemblies, and Final Assemblies.

The following are the relevant Australian Standards, International, National, and state regulations that may be applicable in other countries and the equivalent documentation will need to be substituted:

- Australian Government Hazardous Substances Information system.
- AUS OHS Regulation 2011.
- Australian Government National Occupational Health and Safety Commission, Approved criteria for classifying Hazardous substances NOHSC: 1008 (2004).
- AS 1894 Cryogenic Installations.
- AS 3780–1994 The Storage and Handling of Corrosive Substances.
- AS 4332 Gases in Cylinders.
- AS/NZS 60079.29 Explosive atmospheres. Part 1: Gas detectors-Performance requirements of detectors for flammable gases. Sydney, N.S.W.: SAI Global; 2008.
- AS/NZS 60079.29 Explosive atmospheres. Part 2: Gas detectors-Selection, installation, use, and maintenance of detectors for flammable gases and oxygen. Sydney: SAI Global; 2016.
- AS/NZS 60079.29 Explosive atmospheres. Part 3: Gas detectors-Guidance on functional safety of fixed gas detection systems. Sydney: SAI Global; 2016.
- AS/NZS 60079.29 Explosive atmospheres. Part 4: Gas detectors-Performance requirements of open path detectors for flammable gases. Sydney: SAI Global; 2016.
- ASNZS 2243.10–2004 Safety in laboratories.

Note: This is not a complete listing of the regulations that may apply to any laboratory gas system design; research into the appropriate national regulatory organizations should be carried out.

It may also be necessary to request any laboratory gas specific documentation regarding the design and OHS requirements that the particular facility may have for inclusion.

KEYWORDS

- building management systems
- compressor plant
- sample specification
- vacuum plant

BIBLIOGRAPHY

1. AS 1894–1997, (1997). The storage and handling of non-flammable cryogenic and refrigerated liquids. Homebush, N.S.W.: SAI Global.
2. AS 2896–2011, (2011). Medical gas systems–Installation and testing of non-flammable medical gas pipeline systems (4ᵗʰ edn.). Sydney, N.S.W.: Standards Association of Australia, (2ⁿᵈ edn.). Sydney, NSW: SAI Global.
3. AS/NZS 60079.29, (2008). *Explosive Atmospheres.* Part 1: Gas detectors-performance requirements of detectors for flammable gases. Sydney, N.S.W.: SAI Global.
4. AS/NZS 60079.29, (2016). *Explosive Atmospheres.* Part 2: Gas detectors-selection, installation, use, and maintenance of detectors for flammable gases and oxygen. Sydney: SAI Global.
5. AS/NZS 60079.29, (2016). *Explosive Atmospheres.* Part 3: Gas detectors-guidance on functional safety of fixed gas detection systems. Sydney: SAI Global.
6. AS/NZS 60079.29, (2016). *Explosive Atmospheres.* Part 4: Gas detectors-performance requirements of open path detectors for flammable gases. Sydney: SAI Global.
7. AS 61508–2011, (2011). Part 1: Functional safety of electrical/electronic/programmable electronic safety-related systems. *General Requirements.* Sydney: SAI Global.
8. AS 61508–2011, (2011). Part 2: Functional safety of electrical/electronic/programmable electronic safety-related systems. *Requirements for Electrical/Electronic/ Programmable Electronic Safety-Related Systems.* Sydney: SAI Global.
9. AS 61508–2011, (2011). Part 3: Functional safety of electrical/electronic/programmable electronic safety-related systems. *Software Requirements.* Sydney: SAI Global.
10. AS 61508–2011, (2011). Part 4: Functional safety of electrical/electronic/programmable electronic safety-related systems. *Definitions and Abbreviations.* Sydney: SAI Global.
11. AS 61508–2011, (2011). Part 5: Functional safety of electrical/electronic/programmable electronic safety-related systems. *Examples of Methods for the Determination of Safety Integrity Levels.* Sydney: SAI Global.
12. AS 61508–2011, (2011). Part 6: Functional safety of electrical/electronic/programmable electronic safety-related systems. *Guidelines on the Application of IEC 61508–2 and IEC 61508–3.* Sydney: SAI Global.
13. AS 61508–2011, (2011). Part 7: Functional safety of electrical/electronic/programmable electronic safety-related systems. *Overview of Techniques and Measures.* Sydney: SAI Global.
14. ASTM G88–05, (2005). Standard guide for designing systems for oxygen service (1ˢᵗ edn.). West Conshohocken, PA 19428–2959, United States: ASTM International.
15. ASTM G 93–03, (2003). Standard practice for cleaning methods and cleanliness levels for material and equipment used in oxygen-enriched environments 1 (1ˢᵗ edn.). West Conshohocken, PA 19428–2959, United States: ASTM International.

16. ASTM G 94–05, (2005). Standard guide for evaluating metals for oxygen service (1st edn.). West Conshohocken, PA 19428–2959, United States: ASTM International.

17. ASTM G 127–95, (1995). Standard guide for the selection of cleaning agents for oxygen systems (1st edn.). West Conshohocken, PA 19428–2959, United States: ASTM International.

18. ASTM STP 910, (1986). Flammability and sensitivity of materials in oxygen-enriched atmospheres: In: Benning, (ed.), *American Society for Testing and Materials* (Vol. 2, M.A.). Philadelphia.

19. Brinster, R., & Cross, P., (1972). Effect of copper on the preimplantation mouse embryo. *Nature, 238*(5364), 398–399.

20. CGA V 1:2–13, (2013). Standard for compressed gas cylinder valve outlet and inlet connections, (1st edn.). Chantilly, VA: Compressed Gas Association.

21. *Crane Technical Paper 410 Flow of Fluids Through Valves, Fittings, and Pipe,* (2013), (16th edn.) Stamford, CT: Crane Co.

22. Colebrook, C., (1939). Turbulent flow in pipes, with particular reference to the transition region between the smooth and rough pipe laws. *Journal of the Institution of Civil Engineers, 11*(4).

23. Crowe, C. T., Elger, D. F., & Robertson, J. A., (2005). *Engineering Fluid Mechanics* (8th edn.). John Wiley & Sons.

24. Delmdahl, R., (2015). *Coherent Laser Systems.* Presentation presented at, Göttingen, Germany.

25. *Design & Safety Handbook for Specialty Gas Delivery Systems* (1st edn.), (2017). Plumsteadville, PA: Scott Specialty Gases.

26. Donald, F., Elger, B. A., LeBret, C. T., & Crowe, J. A. R., (2016). *Engineering Fluid Mechanics* (11th edn.). Singapore: Wiley.

27. EN 50271:2010, (2010). *Electrical Apparatus for the Detection and Measurement of Combustible Gases, Toxic Gases or Oxygen–Requirements and Tests for Apparatus Using Software and/or Digital Technologies* (2nd edn.). London: BSI.

28. EN 50402:2017, (2017). Electrical apparatus for the detection and measurement of combustible or toxic gases or vapors or of oxygen–requirements on the functional safety of gas detection systems. *European Committee for Standards–Electrical.*

29. *European Pharmacopoeia,* (2017). Strasbourg: Council of Europe, (9th edn.).

30. *Gascon Systems Auto Change-Over Manifold Information Booklet,* (2013). Melbourne, Australia: Gascon, (1st edn.).

31. Hall, J., & Guyton, A., (2010). *Textbook of Medical Physiology* (12th edn.). Philadelphia, PA: Elsevier Saunders.

32. Schön, H., (2015). Handbook of Purified Gases, (1st edn.). Heidelberg: Springer.

33. Health Technical Memorandum 02–01, (2006). *Medical Gas Pipeline Systems.* Parts 1 and 2. Norwich: Estates and facilities division of the department of health, on behalf of the controller of her majesty's stationery office.

34. Health Technical Memorandum 08–06, (2007). *Pathology Laboratory Gas Systems.* Norwich: Estates and facilities division of the department of health, on behalf of the controller of her majesty's stationery office.

35. Houtermans, M., (2014). *SIL and Functional Safety in a Nutshell* (1st edn.). Zug, Switzerland: Risknowlogy.

36. IEC 61508 2010–04, (2010). Part 1: Functional safety of electronic/electronic/ programmable electronic safety-related systems. *General Requirements*. Geneva, Switzerland: International Electrotechnical Commission.

37. IEC 61508 2010–04, (2010). Part 2: Functional safety of electronic/electronic/ programmable electronic safety-related systems. *Requirements for Electrical/ Electronic/Programmable Electronic Safety-Related Systems*. Geneva, Switzerland: International Electrotechnical Commission.

38. IEC 61508 2010–04, (2010). Part 3: Functional safety of electronic/electronic/ programmable electronic safety-related systems. *Software Requirements*. Geneva, Switzerland: International Electrotechnical Commission.

39. IEC 61508 2010–04, (2010). Part 4: Functional safety of electronic/electronic/ programmable electronic safety-related systems. *Definitions and Abbreviations*. Geneva, Switzerland: International Electrotechnical Commission.

40. IEC 61508 2010–04, (2010). Part 5: Functional safety of electronic/electronic/ programmable electronic safety-related systems. *Examples of Methods for the Determination of Safety Integrity Levels*. Geneva, Switzerland: International Electrotechnical Commission.

41. IEC 61508 2010–04, (2010). Part 6: Functional safety of electronic/electronic/ programmable electronic safety-related systems. *Guidelines on the Application of IEC 61508–2 and IEC 61508–3*. Geneva, Switzerland: International Electrotechnical Commission.

42. IEC 61508 2010–04, (2010). Part 7: Functional safety of electronic/electronic/ programmable electronic safety-related systems. *Overview of Techniques and Measures*. Geneva, Switzerland: International Electrotechnical Commission.

43. IEC 61511–1: 2016–1, (2016). *Functional Safety–Safety Instrumented Systems for the Process Industry Sector*. Part 1: Framework, definitions, system, hardware, and application programming requirements (2nd edn.). Geneva 20, Switzerland: International Electrotechnical Commission.

44. IEC 61511–2016–2, (2003). *Functional Safety–Safety Instrumented Systems for the Process Industry Sector*. Part 2: Guidelines in the application of IEC 61511–1. 1st ed. Geneva 20, Switzerland: International Electrotechnical Commission.

45. IEC 61511–2016–1, (2017). *Functional Safety–Safety Instrumented Systems for the Process Industry Sector*. Part 3: Guidance for the determination of the required safety integrity levels. 2nd ed. Geneva 20, Switzerland: International Electrotechnical Commission.

46. IEC 62061:2005, (2015). *Safety of Machinery–Functional Safety of Safety-Related Electrical, Electronic, and Programmable Electronic Control Systems* (1st edn.). Geneva: International Electrotechnical Committee;.

47. ISA TR84.00.02:2015, (2015). *Safety Integrity Level (47) Verification of Safety Instrumented Functions*. Research Triangle Park, N.C.: ISA-International Society of Automation.

48. ISA TR84.00.02–3:2002, (2002). *Safety Instrumented Functions (SIF)–Safety Integrity Level (SIL) Evaluation Techniques*. Part 3: Determining the SIL of a SIF via fault tree analysis. International Society of Automation.

49. ISO 7396–2011, (2016). *Medical Gas Pipeline Systems*. Part 1: Pipeline systems for compressed medical gases and vacuum (3rd edn.). Geneva Switzerland.

50. ISO 8573–1:2010, (2010). *Compressed Air*. Part 1 contaminants and purity classes (3rd edn.). Geneva, Switzerland: International Organization for Standardization.
51. ISO 8573–1:2010, (2010). *Compressed Air*. Part 2 test methods for oil aerosol content (3rd edn.). Geneva, Switzerland: International Organization for Standardization.
52. ISO 8573–1:2010, (2010). *Compressed Air*. Part 3 test methods for measurement of humidity (3rd edn.). Geneva, Switzerland: International Organization for Standardization.
53. ISO 8573–1:2010, (2010). *Compressed Air*. Part 4 test methods for solid particle content (3rd ed.). Geneva, Switzerland: International Organization for Standardization.
54. ISO 8573–1:2010, (2010). *Compressed Air*. Part 5 test methods for oil vapor and organic solvent content (3rd edn.). Geneva, Switzerland: International Organization for Standardization.
55. ISO 8573–1:2010, (2010). *Compressed Air*. Part 6 test methods for gaseous contaminant content (3rd edn.). Geneva, Switzerland: International Organization for Standardization.
56. ISO 8573–1:2010, (2010). *Compressed Air*. Part 7 test methods for viable microbial contaminant content (3rd edn.). Geneva, Switzerland: International Organization for Standardization.
57. ISO 8573–1:2010, (2010). *Compressed Air*. Part 8 test methods for solid particle content by mass concentration (3rd ed.). Geneva, Switzerland: International Organization for Standardization.
58. ISO 8573–1:2010, (2010). *Compressed Air*. Part 9 test methods for liquid water content (3rd edn.). Geneva, Switzerland: International Organization for Standardization.
59. Loeb, L., (2004). *The Kinetic Theory of Gases*. Mineola, N.Y.: Dover Publications, Chapters 1 and 2.
60. *National Fire Protection Association 55 Compressed Gases and Cryogenic Fluids Code* (2016). Quincy, Mass.: National Fire Protection Association, 1st edn.
61. *National Fire Protection Association 99 Health Care Facilities Code Handbook (Section l.)*, (2018). National Fire Protection Association.
62. Safety Instrumented Systems Verification: Practical Probabilistic Calculations; William M. Goble and Harry Cheddie, PE, CFSE.
63. Yaws, C., (2001). *Matheson Gas Data Book*. Parsippany, NJ: Matheson Tri-Gas.
64. UL 94 standard for tests for flammability of plastic materials for parts in devices and appliances.
65. U.S. Department of Energy, (2008). *Hydrogen Embrittlement of Metals: A Primer for the Failure Analyst*. Aiken, South Carolina 29808.
66. Vidal, F., & Hidalgo, J., (1993). Effect of zinc and copper on preimplantation mouse embryo development in vitro and metallothionein levels. *Zygote [Internet]*, *1*(3), 225–229. Available from: https://www.cambridge.org/core/journals/zygote/article/effect-of-zinc-and-copper-on-preimplantation-mouse-embryo-development-in-vitro-and-metallothionein-levels/5A5083A8BD4A0D95C2EE273D726F F5DE.

INDEX

H

For Product Safety Concerns and Information please contact our EU
representative GPSR@taylorandfrancis.com
Taylor & Francis Verlag GmbH, Kaufingerstraße 24, 80331 München, Germany